Lukas Keller

Molecular Insights into the Eye Evolution of Bivalvian Molluscs

Lukas Keller

Molecular Insights into the Eye Evolution of Bivalvian Molluscs

Isolation and Characterisation of Eye Selector Genes from Arca Noae and Pecten Maximus

Südwestdeutscher Verlag für Hochschulschriften

Impressum/Imprint (nur für Deutschland/ only for Germany)
Bibliografische Information der Deutschen Nationalbibliothek: Die Deutsche Nationalbibliothek verzeichnet diese Publikation in der Deutschen Nationalbibliografie; detaillierte bibliografische Daten sind im Internet über http://dnb.d-nb.de abrufbar.
Alle in diesem Buch genannten Marken und Produktnamen unterliegen warenzeichen-, marken- oder patentrechtlichem Schutz bzw. sind Warenzeichen oder eingetragene Warenzeichen der jeweiligen Inhaber. Die Wiedergabe von Marken, Produktnamen, Gebrauchsnamen, Handelsnamen, Warenbezeichnungen u.s.w. in diesem Werk berechtigt auch ohne besondere Kennzeichnung nicht zu der Annahme, dass solche Namen im Sinne der Warenzeichen- und Markenschutzgesetzgebung als frei zu betrachten wären und daher von jedermann benutzt werden dürften.

Verlag: Südwestdeutscher Verlag für Hochschulschriften Aktiengesellschaft & Co. KG
Dudweiler Landstr. 99, 66123 Saarbrücken, Deutschland
Telefon +49 681 37 20 271-1, Telefax +49 681 37 20 271-0, Email: info@svh-verlag.de
Zugl.: Basel, University of Basel, Diss., 2006

Herstellung in Deutschland:
Schaltungsdienst Lange o.H.G., Berlin
Books on Demand GmbH, Norderstedt
Reha GmbH, Saarbrücken
Amazon Distribution GmbH, Leipzig
ISBN: 978-3-8381-0490-4

Imprint (only for USA, GB)
Bibliographic information published by the Deutsche Nationalbibliothek: The Deutsche Nationalbibliothek lists this publication in the Deutsche Nationalbibliografie; detailed bibliographic data are available in the Internet at http://dnb.d-nb.de.
Any brand names and product names mentioned in this book are subject to trademark, brand or patent protection and are trademarks or registered trademarks of their respective holders. The use of brand names, product names, common names, trade names, product descriptions etc. even without a particular marking in this works is in no way to be construed to mean that such names may be regarded as unrestricted in respect of trademark and brand protection legislation and could thus be used by anyone.

Publisher:
Südwestdeutscher Verlag für Hochschulschriften Aktiengesellschaft & Co. KG
Dudweiler Landstr. 99, 66123 Saarbrücken, Germany
Phone +49 681 37 20 271-1, Fax +49 681 37 20 271-0, Email: info@svh-verlag.de

Copyright © 2009 by the author and Südwestdeutscher Verlag für Hochschulschriften Aktiengesellschaft & Co. KG and licensors
All rights reserved. Saarbrücken 2009

Printed in the U.S.A.
Printed in the U.K. by (see last page)
ISBN: 978-3-8381-0490-4

Table of contents

I. Introduction 5
1. Pecten maximus 5
  1.1 General description of *Pecten maximus* 5
  1.2 Embryogenesis 6
  1.3 The scallop mirror eye 9
  1.4 Gills 12
2. Arca noae 13
  2.1 General description of Arca noae 13
  2.2 Compound eyes in ark clams 15
3. Different eye-types in the animal kingdom 15
  3.1 General comments 15
  3.2 Pigment cup eyes 16
  3.3 Pinhole eyes 17
  3.4 Compound eyes 17
  3.5 The camera type eye 18
4. Eye evolution 20
5. Photoreception 23
  5.1 Two different types of photoreceptor cells 23
  5.2 Usage of distinct phototransduction pathways in rhabdomeric versus ciliary photoreceptors 26
6. Genes involved in the genetic cascade of eye development 27
  6.1 The Pax Gene family 27
  6.2 *Pax6* 30
  6.3 The Six family genes 38
7. The opsin gene family 40

II. Material and Methods 44
1. Molecular methods 44
2. Collection of the animals 44
3. Preparation of genomic DNA 44
4. Isolation of mRNA and cDNA synthesis 45
5. Cryosections 45
6. In situ hybridization protocol 45
7. PCR Protocols 48
  7.1 Degenerated Primers 48
8 RACE PCR 49
  8.1 RACE primers 49
9. Real-time quantitative PCR 51
  9.1. Primers for real-time PCR 51
10. Targeted expression of AnPax6 and PmaPax6 in Drosophila 52
11. Scanning electron microscopy 52

III. Results (Arca) 53
1. Ultrastructure of the Arca noae compound eye 53
2. Arca noae Pax6 (AnPax6) 56
  2.1 Isolation of the *AnPax6* full length cDNA 56
  2.2 Nucleotide and amino acid sequence of *AnPax6* 57
  2.3 Sequence comparison of the Paired domain 58
  2.4 The linker region 59

2.5 Sequence comparison of the homeodomain 60
2.6 Real-time PCR expression analysis of *AnPax6* 61
2.7 *AnPax6* is able to induce ectopic eyes in *Drosophila melanogaster* 62
3. *Arca noae Six1/2 (AnSix1/2)* 64
    3.1 Isolation of the *Arca noae Six1/2 (AnSix1/2)* full-length cDNA 64
    3.2 Nucleotide and deduced amino acid sequence of *AnSix1/2* 64
    3.3 The six domain 65
    3.4 The six homeodomain 66
    3.5 Real-time PCR expression analysis of *Ansix1/2* 67
4. *Arca naoe opsin gene (AnOpsinX)* 70
    4.1 The isolation of *AnOpsinX* full-length cDNA 70
    4.2 Nucleotide and deduced amino acid sequence of *AnOpsinX* 71
    4.3 Structural analysis of AnOpsinX 72
    4.4 Real-time PCR expression analysis of *AnOpsinX* 74

IV. Results (Pecten) 76
1. *Pecten Pax6 (PmaPax6)* 76
    1.1 Isolation of PmaPax6 full length cDNA 76
    1.2 Nucleotide and deduced amino acid sequence of *PmaPax6* full-length cDNA 77
    1.3 Sequence comparison of the paired domain 78
    1.4 The linker region 79
    1.5 Sequence comparison of the homeodomain 79
    1.6 Phylogenetic analysis of bivalvian AnPax6 and PmaPax6 80
    1.7 Real-time PCR expression analysis of *PmaPax6* 81
    1.8 Expression analysis of *PmaPax6* in *Pecten* larvae by whole mount *in situ*
    hybridization 82
    1.9 Targeted expression of PmaPax6 in Drosophila melanogaster 85
2. *Pecten maximus Six1/2 (PmaSix1/2)* 87
    2.1 Isolation of the PmaSix1/2 full-length cDNA 87
    2.2 Nucleotide and deduced amino acid sequence of *PmaSix1/2* 87
    2.3 The six domain 88
    2.4 The six homeodomain 89
    2.5 Phylogenetic analysis of *A. noae* and *P. maximus* Six1/2 91
    2.6 Real-time PCR expression analysis of *PmaSix1/2* 92
    2.7 Expression analysis of *PmaSix1/2* in *Pecten* larvae by whole mount *in situ*
    hybridization 93
3. *Opsin genes in Pecten maximus* 95
    3.1 Isolation of two *Pecten* opsin genes (*PmaGqOpsin* and *PmaOpsinX*) 95
    3.2 Nucleotide and deduced amino acid sequence of *PmaGqOpsin* full-length cDNA .. 96
    3.3 Nucleotide and deduced amino acid sequence of *PmaOpsinX* full-length cDNA 97
    3.4 Structural analysis of PmaGqOpsin 98
    3.5 Structural analysis of PmaOpsinX 100
    3.6 Phylogenetic analysis of *Arca* and *Pecten* opsin genes 102
    3.7 Real-time PCR expression analysis of *PmaGqOpsin* and *PmaOpsinX* 105
    3.8 Expression analysis of *PmaGqOpsin* in *Pecten* eyes by *in situ* hybridization 106

V. Discussion 107
1. *The Arca noae compound eye* 108
2. *The cloning and expression of AnPax6 and PmaPax6* 109
    2.1 Sequence conservation of *AnPax6* and *PmaPax6* 109
    2.2 Expression of *Pax6* in *Arca* and *Pecten* 110
3. *Cloning and expression of AnSix1/2 and PmaSix1/2* 112

   3.1 Sequence conservation of AnSix1/2 and PmaSix1/2 112
   3.2 Expression of *AnSix1/2* and *PmaSix1/2* 113
  ***4.Cloning and expression of three opsin genes in bivalvian molluscs*........................... 115**
  ***5. Conclusions and Perspectives* 117**
VI References.. 120

I. Introduction

1. *Pecten maximus*

1.1 General description of *Pecten maximus*

The bivalve family Pectinidae, also known as scallops, comprise 400 extant species, of which 28 species have been recorded in European waters (Nordsieck, 1969). Scallops are very prominent animals, with many species having commercial importance, mainly because of their flesh, which is much valued as luxury food.

They occur in all seas of the world from polar regions to the tropics. In principal they can be found in all depths, from the intertidal zone down to 7000m or more, however the commercially valuable species occur in the inshore regions of the continental shelves.

Due to its wide distribution and its high market value the "Great scallop" (*Pecten maximus*), also known as "Coquille Saint-Jacques", is commercially the most important scallop in the eastern Atlantic ocean (Figure 1.1.1).

Figure 1.1.1 The valves of *Pecten maximus*. (A) The left concave valve and (B) the right flat valve show the typical "ears" on either side of the apex. Sources: (A) www.unige.ch/sciences/biologie/biani/msg (B) A. Le Maguer-esse/Ifremer (Brest).

Pecten maximus is a large scallop, reaching an average size of 150mm (Minchin, 1978). They are long-living animals, with a life span that can exceed 20 years in extreme cases. *Pecten maximus* is found along the eastern cost of the North Atlantic from northern Norway south to the Iberian peninsula (Tebble, 1966) and has also been reported to occur in West Africa, the Azores, Canary Islands and Madiera (Mason, 1983). They were reported to even extend a short distance into the Mediterranean sea as far as in the province of Malaga (Cano and Garcia, 1985). Further east it is replaced by the closely related *Pecten jacobaeus*, which occurs throughout the Mediterranean (Piccinetti et al., 1986).

Pecten maximus has unequal valves, with an upper (right) valve that is flat and a lower (left) that is strongly concave and generally overlaps the left valve at its margin (Figure 1.1.1). Both valves bear 15-17 broad, radiating ribs and numerous concentric corrugations with fine striae. The right valve is commonly off-white, yellowish or bright brown, the left valve is commonly reddish brown but may vary from light pink to almost black.

They prefer shallow habitats at a depth of 20-45m, generally on sandy bottoms, fine gravel or sandy gravel sometimes with an admixture of mud (Mason, 1983).

In contrast to most other bivalves which are dioecious, scallops are hermaphrodites. The mature gonad contains a proximal, creamy-coloured testis and a distal, orange-coloured ovary of approximately equal size. In the case of *Pecten maximus* the gametes are released simultaneously during spawning.

1.2 Embryogenesis

Development of scallops was heavily investigated, mainly for the need to improve culturing conditions and to develop methods to replenish the rapid shrinking natural population due to excessive exploitation. Most of the information about development were sampled so far from commercially important species and under hatchery conditions. In contrast, there are only few studies on development under natural conditions, because of the difficulties to identify and monitor the tiny planktonic larvae within the water mass. During spawning the eggs and the sperms are released simultaneously into the sea and fertilization occurs externally (Figure 1.2.1). At the time of sperm penetration the eggs are at metaphase I stage of meiosis. (Gruffyd and Beaumont, 1970). Division is spiral, complete and hetero-quadrantal and cleavage leads to an immotile stereoblastula. (Kulikova and Tabunkov, 1974; Malakhov and Medvedeva, 1986; Tanaka, 1984). Subsequent gastrulation occurs by epiboly and invagination (Drew, 1906; Fullarton, 1896; Gutsell, 1930; Hodgson and Burke, 1988) that leads to the first stage in scallop development which has cilia and therefore is motile. However, the movements are yet undirected and consist of rolling and spinning. Further development of the spherical gastrula leads to a trochophora larva. Depending on the

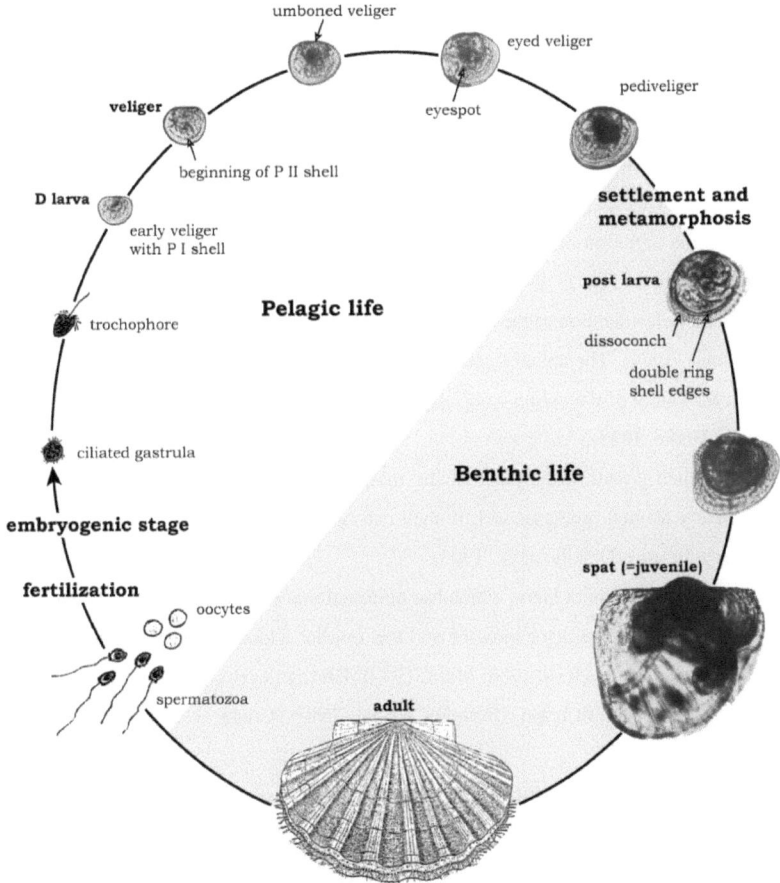

Figure 1.2.1 Diagrammatic representation of the *Pecten maximus* life cycle. PI and PII, prodissoconch I and II, respectively. From (Le Pennec et al., 2003).

temperature and other environmental conditions, the development from fertilization to the trochophora larval stage takes approximately 24 hours (Comely, 1972). At this stage the first sensory organs arise. A prominent feature is the sensory apical flagellum, which however is lost within a day or two. At the later veliger stage it remains as a short apical tuft. The apical flagellum is in fact a bundle of up to 50 long cilia adhering to one another to form a whip-like structure (Bellolio et al., 1993). Its precise function so far is enigmatic and still awaits its thorough investigation. A plausible possibility is that it may play a role in chemoreception or may work as a mechanoreceptor. On the dorsal surface of the trochophore, a surface infolding develops which give rise to the shell gland (Bellolio et al., 1993; Casse et al., 1998; Fullarton, 1896; Malakhov and Medvedeva, 1986; Sastry, 1965).

The transition from trochophore to veliger larva starts with the development of the most characteristic organ of bivalve veliger, the velum (Gruffyd and Beaumont, 1972; Malakhov and Medvedeva, 1986; Sastry, 1965). This ciliated organ allows locomotion and then progressively enables the capture of food particles. A further sign for the transition to veliger larva is the initiation of the Prodissoconch I shell secretion, which leads the D-shaped early veliger (Figure 1.2.1); (Casse et al., 1998; Malakhov and Medvedeva, 1986). Development from fertilization to the first D-veliger stage takes approximately 48h (Comely, 1972; Le Pennec, 1974). The apical flagellum of the trochophora is now transformed into an apical tuft in the center of the velum composed of cilia which do not adhere together anymore (Hodgson and Burke, 1988).

Further shell-growth takes place at the margin of the prodissoconch I shell (Figure 1.2.1). The newly formed prodissoconch II shell can be distinguished from the former by its concentric growth rings (Bellolio et al., 1993).

Unlike the trochophora larva, which has no functional muscles, the early veliger starts to build up the first velar retractor muscles and the anterior adductor muscle (Bellolio et al., 1993; Malakhov and Medvedeva, 1986; Maru, 1972). During development the number of the velum retractor muscles increase (Bellolio et al., 1993; Cragg, 1985; Malakhov and Medvedeva, 1986).

Further development of the digestive tract, comprised of an archenteron and the blastopore at the trochophora stage, leads to the second opening, the anus (Hodgson and Burke, 1988; Malakhov and Medvedeva, 1986). The mouth is situated at the posterior end of the velum, opening into a straight ciliated cylindrical oesophagus with actively beating cilia (Beaumont et al., 1987) that leads to the stomach. The thin intestine is initially straight but develops into one and then two loops (Bower and Meyer, 1990). The anus is located close to the hinge line in the posterior body wall.

A posterior adductor muscle develops by the late veliger or pediveliger stage, consisting of two columns. In contrast, the anterior adductor muscle consists of only a single column (Bellolio et al., 1993; Cragg, 1985). At the same time a pair of eye spots is found that can be seen through the transparent shell, appearing approximately in the centre of the valve when the larva is viewed from the side (Figure 1.2.1). They consist of one cell with pigmented granules forming an anteriorly directed cup and another, yet uncharacterized cell within the cup, most probably a photoreceptor cell (Hodgson and Burke, 1988).

Another prominent feature of bivalves that develops at the late veliger stage are the gills. Around the second week of the pelagic veliger phase, the gill anlagen differentiate from the general external epithelium (Beninger et al., 1994). The cells of the anlagen are first symmetrically positioned to the left and the right base of the foot and later form distinct buds, or primordia.

At the end of the larval life, pediveligers initiate settlement behaviour. This process leads to a loss of some larval organs and changes in the nature of shell secretion. The primary organs which are lost are the velum, the velar retractor muscles and the anterior adductor muscle (Cragg, 1985). The gills now increase in length and organize into a row of straight gill filaments as other buds form and then grow in a posterior-anterior sequence, increasing the number of filaments. (Beninger et al., 1994). The gills are nonfunctional in the primordial bud stage and become functional in a gradual manner during metamorphosis.

1.3 The scallop mirror eye

Since bivalves are commonly known as animals that lost their sense organs and their cerebral nervous system during evolution, it seems somewhat counterintuitive that bivalves may have eyes, all the more as eyes are mostly found in cerebral regions in other animals. However, a variety of different eye-types are found in bivalves, ranging from primitive pit eyes to elaborated camera-type eyes as for example found in the heartshell *Cardium*.

The eyes of scallops have attracted anatomists since the late 18th century, mainly because of their resemblance to the camera eyes of vertebrates. Dakin, at the beginning of the 20th first described the essential morphological features of the scallop eye and clarified the former idea of a camera eye.

The eyes of scallops are located at the tips of short stalks that peep out from the middle fold of the mantle margin (Figure 1.3.1A). Generally several dozens of mirror eyes are found in a single animal, in some rare cases more than 100 (Dakin, 1910) but varies from individual to individual. No correlation has been found between the size of the shell and the number of eyes. Eyes are found in association with both, the upper (left) and lower valves, although they occur in greater number and size on the upper mantle (Gutsell, 1930). Moreover, eyes of smaller sizes are spaced irregularly among those of full size (Dakin, 1910). The epithelial cells of the eye are heavily pigmented on the lateral sides (Figure 1.3.1C). Only the most distal part has a clear cornea that enables the light-rays to enter. A lens comprising irregular cells lies just beneath the cornea. At the back of the eye lies the retina, the reflecting argentea (also called tapetum) and a pigmented layer.

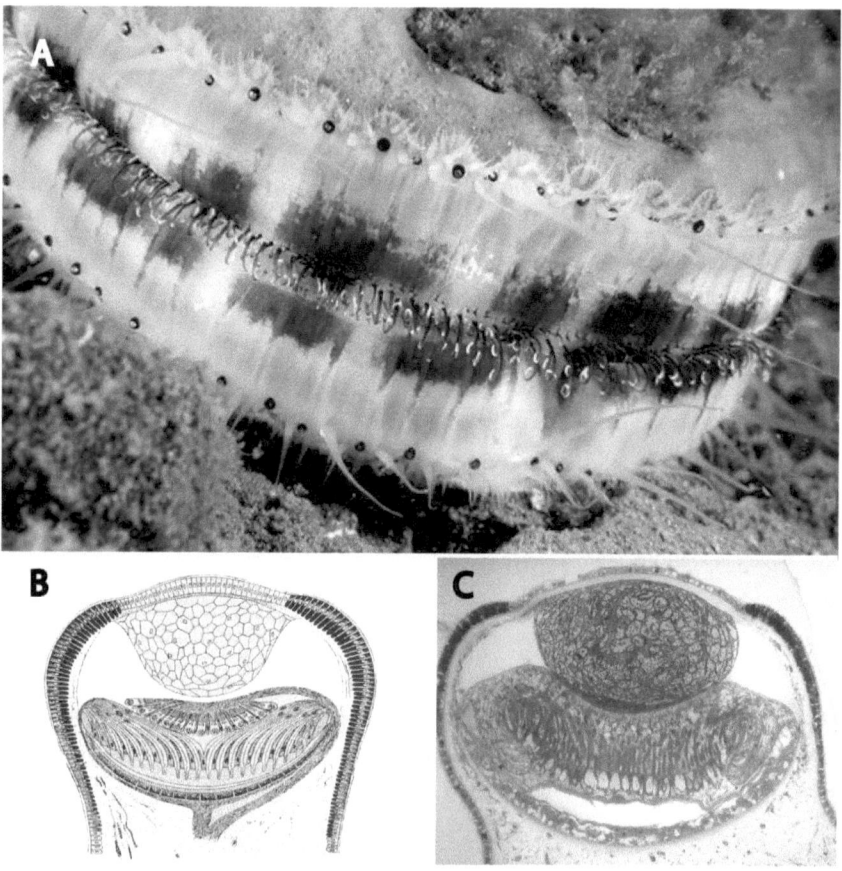

Figure 1.3.1 The mirror eyes of *Pecten maximus*. (A) The eyes are located at the mantle margin of the animal and have a shiny appearance because of the mirror in the back of the eye that reflects the light back (Courtesy of Ron Offermans). (B) Schematic representation of a scallop mirror eye (after Küpfer). (B) Paraffin section through a *Pecten* eye stained with methylene blue (Sauder und Keller).

An extraordinary feature of the scallop retina is its organisation into two distinct layers of retinal cells. The proximal retinal layer is build up by rhabdomeric microvillar photoreceptors, whereas the distal retinal layer comprises ciliary photoreceptor cells (Figure 1.3.2A). The sensory region of the proximal photoreceptors are oriented towards the argentea, whereas the ones of the distal photoreceptors face the lens and cornea (Barber et al., 1967).

Therefore, in respect of the argentea, the proximal photoreceptors are everted whereas the distal ones are inverted.

The axons from the proximal photoreceptor layer extend around the sides of the retina (Figure 1.3.2A). Axons from the distal photoreceptor layer run laterally to the cilia, pass in front of the retina and collect at the retinal margin at the lateral side to form the distal branch of the optic nerve. Thus, each retinal layer forms a separate nerve branch which extends and joins the other branch 1 to 2mm proximally from the eye capsule to form the optic nerve (Barber et al., 1967; Hartline, 1938; Miller, 1958).

The argentea, the reflecting layer behind the eye, is responsible for the bright iridescence of the pupil. It is build up by a single layer of cells containing an array of flattened, membrane bound guanine crystals. Forming a precise hemispheric shape it acts as a perfect concave mirror reflecting the light back to the retina with a focal length of approximately 200µm (Land, 1965). The focal length of the mirror is almost precisely the distance to the distal retinal layer. Therefore, the inverted image is formed at the distal retina. The lens lies in contact with the retina (Figure 1.3.1B and C) and hence has no focusing function since the focal length of the lens has been shown to be 1.5mm, which lies far behind the eye. Thus, the only plausible function of the lens is to correct the spherical aberration of the argentea.

Another difference of the two retinal layers is found in their electrophysiology and the physiological behaviour. The ciliary photoreceptors of the distal retinal layer hyperpolarize in respond to light impulses, whereas light stimulation of the proximal rhabdomeric photoreceptors leads to a depolarization (Figure 1.3.2B). Studies on the physiological properties of photoreceptors demonstrated that the distal retinal layer responds when light is turned off or is reduced in intensity. In contrary, the proximal retinal layer responds when the eye is illuminated (Hartline, 1938; Land, 1966). The functional importance is that only the distal "off" receptors lie in the plane of focus of the argentea. Hence it is only the distal retina which responds to movements of the visual field that lead to changes in light intensities. It is thought therefore, that the distal retina works as a shadow detector. Experimental studies by stimulating eyes with regular patterns of light/dark stripes showed that scallops can detect movements in the environment at a distance greater than that required to cast a direct shadow on the animal. This is undoubtedly a distinct advantage that allows to react appropriately to fast moving predators.

The "On" receptors of the proximal retinal layer react to changes in overall light intensities. The function of these photoreceptors are less well understood, however it is

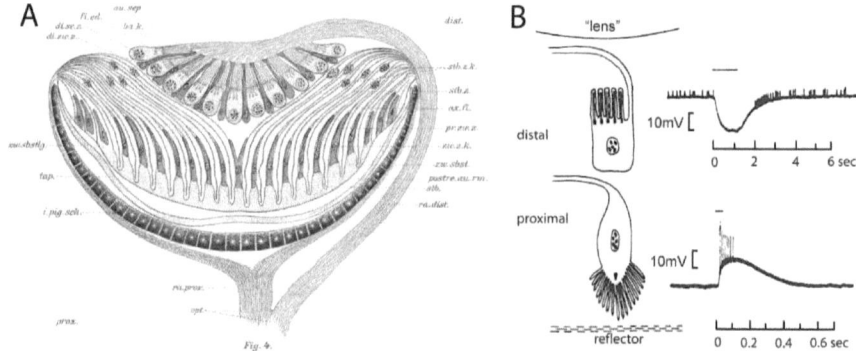

Figure 1.3.2 The *Pecten* retina and its physiological behavour. (A) The proximal retina is build up by rhabdomeric photoreceptor cells, whereas the distal retina has ciliary photoreceptor cells (after Küpfer). (B) The distal ciliary photoreceptor cells hyperpolarize when stimulated, whereas the proximal rhabdomeric photoreceptor cells depolarize in respond to light. From (Land, 1965).

believed to serve as a detection of absolute levels of light intensities and might be useful for migration and habitat selection.

1.4 Gills

Scallops have heterorhabdic (of different size), plicate gills. The W-shaped left and right gills are composed of a series of two different types of filaments, suspended from the gills axis in a plicate fashion. The gill filaments gradually decrease in length toward the anterior and posterior extremities. In the anteriormost region of the gills this shortening of the filament results in the convergence of the dorsal feeding tracts with the oral groove at the base of each pair of labial palp. The gill filaments are essentially hollow tubes within which the haemolymph circulates.

Interestingly, a sensory organ, the so called osphradium, has been found on the gill axis (Figure 1.4.1 A) (Haszprunar, 1987a). This sensory epithelium is situated on the mid-portion of the gill axis and extends from the anterior region of the gill for approximately four-fifth of the length of the gill axis. There is little known about its function, but it has been proposed that it may serve as an organ for chemoreception (Haszprunar, 1987a; Haszprunar, 1987b), which might be associated with detecting spawning cues of conspecies and of gamete releasing signals (Beninger and Donval, 1995; Haszprunar, 1987a; Haszprunar, 1987b).

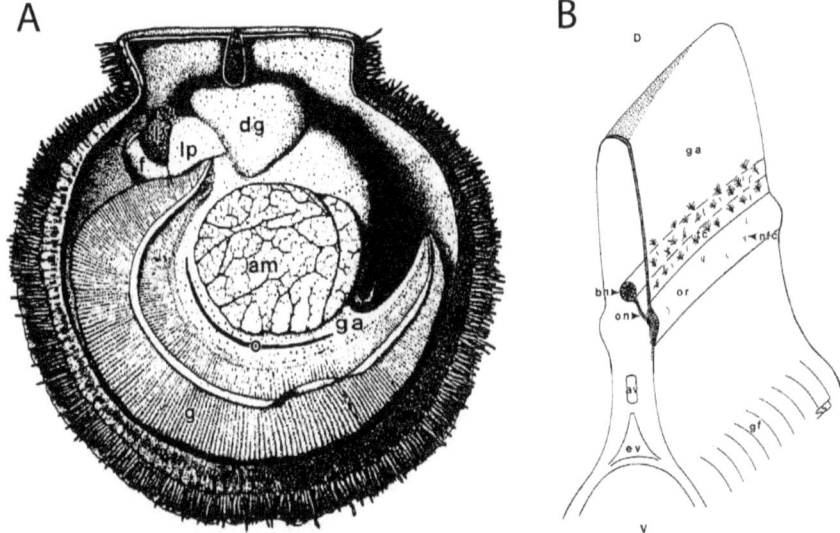

Fig 1.4.1 Anatomical localization of the gills and the osphradium. (A) Schematic drawing of *Placopecten magellanicus* with the left valve removed. Note the location of the ospradium (o) along the lateralmost margin of the gill axis. *am* Adductor muscle, *dg* digestive gland, *f* foot *g* gill, *ga* gill axis, *l* lips, *lp* labial palp. (B) Schematic representation of anatomical relationships and innervation of the osphradium. av Afferent branchial vessel, bn branchial nerve, D dorsal orientation, ev efferent branchial vessels, ga gill axis, gf, gill filaments, cilium of free nerve fiber on osphradial ridge, or osphradial ridge, on one of the osphradial nerves, tf tuft cilia, V ventral orientation. Figure from (Beninger and Donval, 1995).

2. Arca noae

2.1 General description of Arca noae

Much less is known about the ark clam *Arca noae* (Linnaeus 1758). They are found in the eastern Atlantic Ocean, the Mediterranean and Black Sea and moreover in the Caribbean Sea (Nordsieck, 1969). *Arca noae* is firmly attached by a solid byssus to rocks or shells and may occur down to depths of 100m. It has been reported that they are able to slowly move over rocks at night and that solitary byssus may be seen abandoned on the previous anchoring point(Marin and Lopez Belluga, 2004). They can reach a maximum length of 70-90 mm and may live up to 15 years (Hrs-Brenko and Legac, 1996; Poppe and Goto, 2000).

Figure 2.1.1 *Arca noae* (A) The shells of *Arca noae* have a typical ark shaped appearance, hence the name. (Source: http://digilander.libero.it/conchigliaveneziane/bivalvi/specie/ArcaNoae.htm). (B) *Arca noae* is often covered by a the red sponge called *Crambe crambe* (Courtesy of Miquel Pontes).

Arca noae is characterized by boat-shaped shells (Figure 2.1.1A), hence their name, which are sculptured by radial ribs and are bright brown coloured with dark markings. Commonly, *Arca noae* are associated with a red demosponge (*Crambe crambe*), one of the most widespread littoral sponge species in the western Mediterranean (Figure 2.1.1B) (Becerro et al., 1994). Association of *Arca* with *Crambe*, which grows on the shells of the animal, was shown to decrease its predation (Marin and Lopez Belluga, 2004). Indeed, *Crambe* was shown to contain a group of potent cytotoxic and antiviral secondary metabolits called crambescidins (Jares-Erijman et al., 1991).

There are some reports of commercial exploitation in the Adriatic Sea, particularly in Croatia, but fishing may also occur on the Mediterranean cost of Southern France (Benović, 1997). *Arca* are harvested primarily by divers and are generally sold at local markets. However, the commercial value of *Arca* is very low and there is no report of any launched artificial culturing. This is probably the reason why little or nothing is known about *A. noae* development. Because trochophore and veliger larvae are very small and because of the difficulties to distinguish veligers of different species, it is almost impossible to study their development in their natural environment. There are so far no studies about *Arca noae* development and little is known about the seasonal cycle of gametogenesis and spawning periodicity.

2.2 Compound eyes in ark clams

Ark clams may have up to 300 compound eyes, each composed of several dozens ommatidia, located at the mantle margin. In addition, several hundreds of pigment cup eyes are scattered between the compound eyes (Janssen, 1991; Morton, 1987; Patten, 1886). In bivalves, the mantle edge is divided into three folds, the middle of which usually carries the eyes. In contrast, the compound eyes of ark clams are found on the outer fold of the mantle margin, a unique feature of the Arcacea family (Waller, 1980). Most of the eyes are located in the anterior and posterior part of the mantle edge, whereas in the middle part only few eyes are found. The eyes vary considerably in size, with smaller eyes intermingled between larger eyes.

Unlike arthropods which have eight to nine photoreceptors per ommatidium, ark clams have only one (Eakin, 1963). Another striking difference between compound eyes of arthropods and ark clams is the lack of a focusing device (Nilsson, 1994).

Originally it was thought that the photoreceptor cells of ark clams are rhabdomeric. However, subsequent studies showed that compound eyes of ark clams have ciliary photoreceptor cells. (Levi and Levi, 1971; Nilsson, 1994).

3. Different eye-types in the animal kingdom

3.1 General comments

Nature invented an enormous range of eye-types during the course of evolution (Figure 3.1.1). Of the approximately 33 animal phyla, about a third have no specialized organs for light-detecting, whereas the remaining two thirds have light detecting organs (Land and Nilsson, 2002). As animals are found in different habitats and adopt variable lifestyles, their eyes too had to adapt to the appropriate environment. Aquatic animals, for example, face other optical problems than terrestrial animals. The cornea of aquatic animals is nothing more than a tough transparent membrane which protects the surface of the eyeball but has little or no optical effect, because the fluid has the same refractive index on both sides of the cornea. In land-living animals, however, the front surface of the cornea is in air and becomes a focussing device.

Another example are nocturnal animals, which face the problem to catch enough photons. These animals usually have very large eyes, as seen for example in deep-sea animals where little or no light penetrates. A structure often found in deep-sea animals or animals that are active at night is the "tapetum lucidum", a mirror behind the retina. The function of this structure is to reflect the light already focused by the lens back to the retina, giving the retina a second chance to capture the photons missed by the first pass. These examples are surely features of already quite elaborated eyes and there are of course circumstances where much more simple eyes fulfill the needs of their bearers.

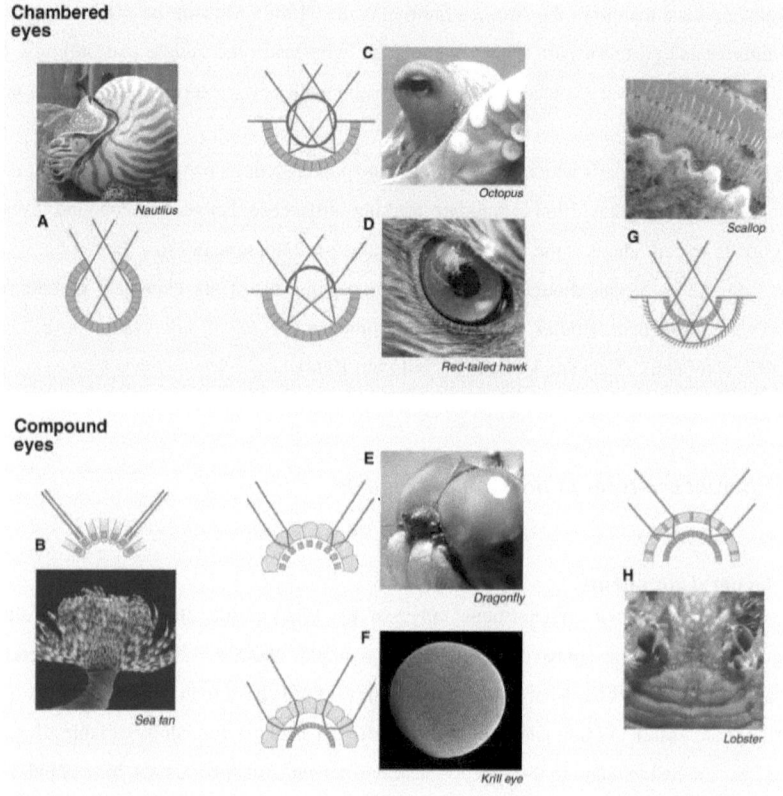

Figure 3.1.1 Various eye-types found in the animal kingdom. Eyes can be generally grouped into either chambered eyes or compound eyes. For more details see (Fernald, 2006)

3.2 Pigment cup eyes

The pigment cup eye consists of just two cell types, photoreceptor cells and pigment cells. In the simplest case, a pit eye is build up by a single photoreceptor cell and one shielding pigment cell. Such primitive two-celled eyes are found, for example, in the japanese

16

planarian *Polycelis auricularia* (Gehring, 2004) and many lophotrochozoan larvae e.g. the trematode *Multicotyle purvisi* (Rhode and Watson, 1991) or the polychaete *Platynereis dumerilii* (Arendt et al., 2002). Usually however, pigment cup eyes are composed of multiple photoreceptor and pigment cells and various cellular arrangements are found in nature. Several photoreceptor cells can share a single cup-shaped pigment cell as in the turbellarian flatworm *Bdellocephala brunnea* (Kuchiiwa et al., 1991) or they are shielded by a pigment cup consisting of multiple pigment cells. The photoreceptors may have an inverted orientation where the photoreceptive organelle is orientated towards the cavity of the pigment cell or an everted orientation pointing towards the light source.

The pigment cup eye has some ability to compare light-intensities in different directions, though the image forming power is very poor.

3.3 Pinhole eyes

What distinguishes pinhole eyes from pigment cup eyes is their size. Most of the pigment cup eyes are just a fraction of a millimeter in diameter, with a few dozens of photoreceptors. A way to improve the performance of an eye is to make the eye bigger and the aperture smaller. In *Nautilus*, the most prominent representative of the pinhole eye, the eyes are nearly a centimeter in diameter, comparable in size of the lens-containing eyes of *Octopus* (Land and Nilsson, 2002; Muntz and Raj, 1984). Giant clams (*Tridacna*) also have pinhole eyes around their mantle margin allowing them to detect moving objects.

The lack of a focusing device is the weak point of the pinhole eye design. In the case of the *Nautilus* pinhole eye, the resolution can be improved by decreasing the size of the pupil. However, this is far from being ideal, because small aperture reduces the amount of light reaching the photoreceptive field leading to an image which is very dim. In contrary, increasing the aperture results in a loss of resolution. Therefore, the image formed by a pinhole eye is either blurred or very dim.

3.4 Compound eyes

Compound eyes are by far the most popular visual system regarding the large number of species that possess them. This type of eye is widely used by arthropods, predominantly in insects, but is also found in some representatives of the lophotrochozoan clade as for example in ark shells and sabellid tubeworms (Nilsson, 1994).

It is thought that the compound eyes arose around the time of the Cambrian explosion, some 530 million years ago. Indeed, in some well preserved trilobite fossils it is possible to see the facets of the compound eye.

Compound eyes differ from the more primitive pigment cup eyes by having a lens associated with each photoreceptor or more, usually a cluster made up of eight to nine photoreceptors, forming a unit called ommatidium. The lens substantially improves vision by gathering more light to stimulate the photoreceptor cells. Moreover they function to define the visual field of each ommatidium (Nilsson, 1989). Although each ommatidium forms its own tiny inverted image, the overall image projecting to the brain is erect and formed from the apposed image of the visual field of individual ommatidia. Hence the reason why this type of compound eye is called apposition eye.

A weak point of the compound eye design is the small size of the optical elements, which are typically around 25μm in diameter (Land and Nilsson, 2002). The problem lies in the universal rule of the optics that the smaller the diameter of an aperture, the larger is the interference pattern in the image produced by light from a point in object space. Therefore, the resolution a compound eye can provide is limited by diffraction. The constraints become clearer if we imagine a compound eye with a resolution comparable to the human eye. In such a case, the compound eye would need to have a diameter of one meter, a dimension hardly feasible for any of the numerous insect species (Kirschfeld, 1976).

In many nocturnal insects and some crustaceans nature invented an optical system to increase the sensitivity of the compound eye (Nilsson, 1989). This improvement is achieved by using an optical design where the light reaching the photoreceptors comes not from only one optical element (the lens) but from many. Such compound eyes are known as superposition eyes, since light from many elements are superimposed. They differ from apposition eyes being less obviously divided into ommatidia and having a single, deep-lying retinal layer separated from the optical elements by a transparent region.

3.5 The camera type eye

The major difference between the pinhole eye and the camera-type eye is the acquisition of a lens of the latter. As discussed in the previous paragraph, the optical design of a pinhole eye either leads to a very dim or a blurred image. To circumvent this disadvantage, nature invented a focusing device, the lens, which enables the eye to increase its resolution power without decreasing the amount of incident light. Eyes of this constructions are found in vertebrates, in cephalopods other than *Nautilus*, but also in some gastropod molluscs, some annelid worms and at least in one copepod (Land, 1984). The lens is generally made of proteins that have a higher refractive index than the surrounding medium. In aquatic animals, the lens is usually spherical because a sphere provides the shortest focal length and in addition provides the most compact form. However, spherical lenses constitute a serious problem. Light-rays which hit the outer region of the lens are bent too much and focus closer to the lens than rays striking closer to the centre of the lens, a phenomenon called spherical aberration. In fishes, this problem is solved by forming a lens with gradient of refractive index with the highest in the centre and the lowest in the periphery (Jagger, 1992). In land-dwelling vertebrates the transparent cornea

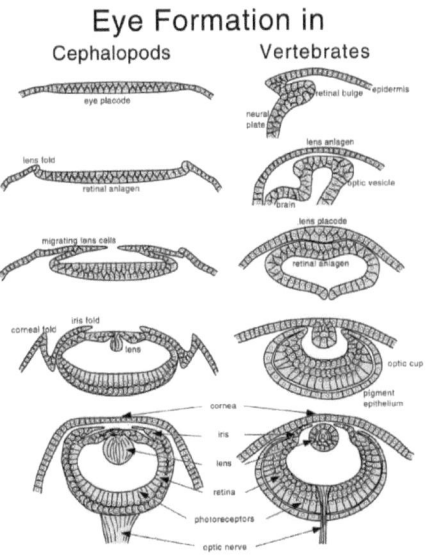

Figure 3.5.1 Schematic diagram of cephalopod eye development (Left) and vertebrate eye development (Right). Development proceeds from top to bottom. The cephalopod eye forms from an epidermal placode through a series of successive infoldings, while the vertebrate eye emerges from the neural plate and induces the overlying epidermis to form the lens. From (Harris, 1997).

becomes a refractive device too, because on land the lens is exposed to air on one side and to water on the other. Just adding the new optical power of the cornea to that of the spherical lens would result in an eye forming the image far in front of the retina. How is the problem solved then? To loose the protecting cornea is probably not a very save solution, hence it is the lens which has to be rejected or modified. Indeed, the lens of land vertebrates retain their lenses but with much weaker focusing power. In humans, for example, the cornea is accountable for two thirds of the light-bending activity (Charman, 1991). During this adaption, the lens become more a device for accommodation (focusing to different distances) than for providing focusing power. As homogeneous lenses, the cornea too is not immune of spherical aberration. In human, for example, this is counteracted by a slightly dome-shaped surface of the lens.

Worth mentioning is the case of the cephalopod eye versus the vertebrate eye, since this was a textbook example for convergent evolution (Figure 3.5.1). Although they are extraordinary similar in design, they differ in their ultrastructure as well as in their ontology. Anatomically, there are two main differences. In the first place they use different types of photoreceptor cells. Whereas cephalopods use rhabdomeric photoreceptor cells, the vertebrate retina consists of ciliary photoreceptor cells. Secondly, the photoreceptor cells of cephalopods are orientated towards the light-source, representing the everted configuration, whereas photoreceptor cells of vertebrates are inverted, pointing away from the light-source. Also the embryonic origins of the eye differ remarkably. Vertebrate eye development starts with a neuroectodermal outgrowth from the lateral forebrain giving rise to an optic vesicle (Figure 3.5.1). The optic vesicle comes into close contact with the overlying surface ectoderm and induces the formation of the lens placode. Further development leads then to the invagination of the optic vesicle, forming two retinal layers, an inner neuroretinal layer and an outer retinal pigmented epithelium. Concomitantly, the lens placode develops into a lens and the covering epidermis into a transparent cornea. In cephalopods however, all three structures, the retina, the lens and the cornea develop from the surface ectoderm. Initially, the surface forms an eye placode of thickened cells, which then invaginates to form the retina. The lens, iris and cornea form from successive folds of the ectoderm that encircle the developing eye. In contrast to vertebrates, the lens of cephalopods are acellular and develop as two approximately hemispheric halves from two separate ectodermal sources (Sivak et al., 1994).

Another important optical system found in nature is the mirror eye. Since this eye-type was extensively discussed for scallops in a previous section, a further description of the mirror eye is unnecessary.

4. Eye evolution

In modern animals various eye-types with intriguing complexity and amazing acuity are found. For a long time it was a complete enigma how such a perfect device as the vertebrate eye, which is capable to adjust the focus to different distances and to adopt to different light intensities and furthermore is able to correct spherical and chromatic aberration, could have evolved. Already Darwin, in his seminal work "The origin of species", admitted that the idea of an organ as perfect as the eye could have been formed by natural selection seems plainly counter-intuitive. Each part of the eye, as for example the lens or the retina, are essential to enable proper vision. How is it possible then to explain eye evolution by natural selection if each of its components are nonredundantly needed for accurate function? Darwin, completely aware of these troubles, proposed a genuine hypothesis. He assumed a primitive and yet imperfect eye, a prototype eye, on which natural selection could act (Figure 4.1). This prototype eye, Darwin proposed, consists of at least two cells: a photosensitive cell (photoreceptor cell) and a pigment cell that shields the photoreceptor cell from one side. Indeed such two-celled eyes were found, for example, in trochophora larvae and in planarians (Figure 4.1) (Arendt and Wittbrodt, 2001; Gehring and Ikeo, 1999).

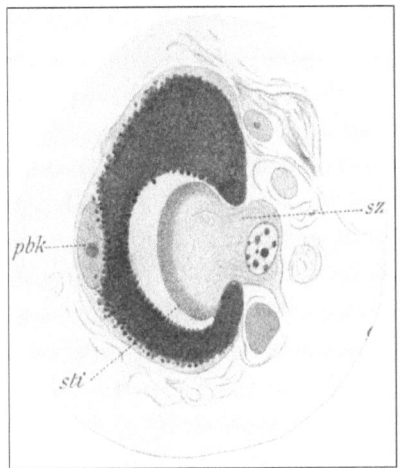

Fig 4.1 Histological section through a prototype like eye (Planaria torva) consisting of three photoreceptor cells and one shielding pigment cell. *sti* microvilli, *sz* photoreceptor cell, *pbk* pigment cell nucleus. After (Hesse, 1897).

From this prototype then, more sophisticated eye could have evolved in a gradual manner by variation and natural selection. However the prototype eye itself can not be explained by natural selection, since natural selection can only work once the eye functions at least partially. Therefore the origin of the prototype eye must have been a stochastically very improbable event. But what was the driving force in evolution to generate a mechanism for light perception? There is no necessary need for light or visual perception to interact with the environment. There is a vast array of sensory perceptions in animals ranging from olfactory perception to the ability to sense electric fields or to notice the terrestrial magnetic field. However, there must have been a selective advantage during evolution for light perception. Almost every, if not all organisms are known to have the ability of light perception, from bacteria to protists up to higher metazoans. Gehring and Roshbach (Gehring and Rosbash, 2003) proposed that the capability to detect light must have been a selective advantage in the early phases of evolution. Geological studies provide evidence that in precambrian times the atmosphere contained little oxygen and a protecting ozone layer was lacking. Therefore organisms were exposed to heavy doses of destructive UV irradiation during daytime. The strong selective pressure by UV irradiation most probably drove the evolution of specialized photoreceptors. Since life originated in the oceans, Gehring and Roshbach suggest that the new feature to sense light intensity enabled organisms to avoid UV irradiation by descending in the oceans. In a nutshell, the early

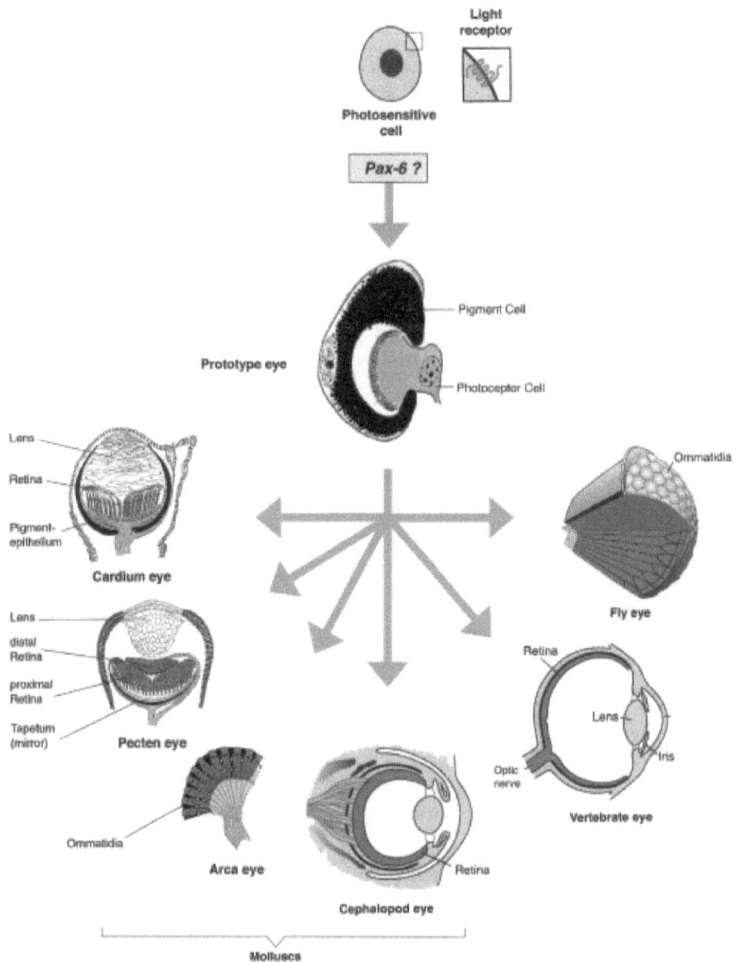

Figure 4.2 Monophyletic evolution of various eye-types starting from a *Pax6*-dependent Darwinian prototype eye consisting of one photoreceptor cell and a shielding pigment cell. From (Gehring and Ikeo, 1999).

environmental conditions on earth exerted a strong selective pressure in favour for the evolution of a light sensing device in living organisms. As evolution went on, the new acquirement of light perception was gradually developed to higher levels of complexity and eventually gave rise to sophisticated eyes allowing spatial vision. Nature produced a plethora of eyes in metazoans, at various locations and of breathtaking morphological diversity. The wealth of eye diversity caused Neo-Darwinist to propose, most prominently Salvini-Plawen and Mayr (1961), that photoreceptor organs originated independently at least in 40, but possible up to 65 or more different phyletic lines. However their conclusion is purely based on comparative morphological and ultrastructural reasons and excludes critical facts that argue rather for a monophyletic origin of the eye. The most striking evidence for a monophyletic origin is found on the molecular level. An important argument for a monophyletic origin is, for example, the observation that all metazoans share the same visual pigment, rhodopsin. But even on pure morphological reasons it is highly unlikely that eyes evolved 40 to 60 times independently in different phyletic lines. The finding that within a single phyletic class of bivalvian molluscs all major eye-types are represented (compound eyes in ark shells, camera-type eyes in cockles (*Cardium*) and mirror eyes in scallops) makes it highly improbable that all these eye-types evolved independently in the bivalvian class. Given that new formations are stochastically rare events, it seems more plausible to propose that these eyes arose by divergent evolution from a common ancestor eye than to argue for the rather improbable event that they emerged independently. The most powerful evidence for a monophyletic origin of the eyes is found at the level of specifying transcription factors. One of these, the gene *Pax6*, has been shown to play a very important role in developing eyes throughout the animal kingdom. The wide use of *Pax6* as a master control gene for eye development can be best understood as a reflection of a very ancient *Pax6* involvement for the specification of a pre-bilaterian photoreceptor cell precursor (Figure 4.2) (Gehring and Ikeo, 1999; Pichaud and Desplan, 2002).

5. Photoreception

5.1 Two different types of photoreceptor cells

All photoreceptor cells face the problem to store as many photopigment molecules as possible to gain optimal light-sensitivity. Since the photopigment, rhodopsin, is a membrane protein traversing the lipid bilayer seven times, storage can be enhanced by enlarging the membrane surface. Indeed, photoreceptor cells do so by local in- or outfolding of their membrane surface, forming a light-sensitive organelle. From ultrastructural studies it is long known that nature found two ways to do so. One strategy is to fold the apical cell surface into numerous microvilli forming a structure called rhabdom found in rhabdomeric photoreceptor

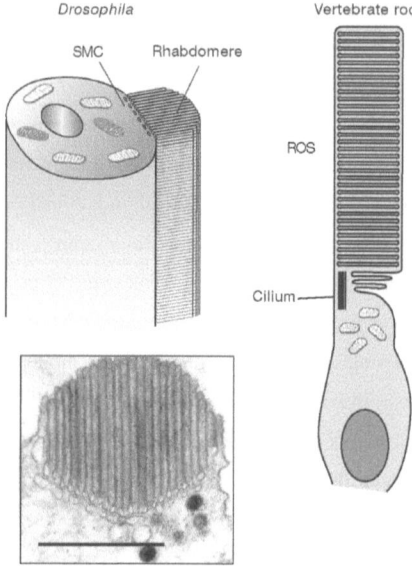

Figure 5.1.1 Ciliary and rhabdomeric photoreceptor structures. The photoreceptive membrane of rhabdomeric photoreceptors is build up by tightly packed tubular microvilli forming a rhabdomere. Vertebrate rod outer segments (ROS) contain stacks of membranous discs and are connected to the cell body by a cilium. SMC: submicrovillar cisternae. From (Hardie and Raghu, 2001b).

cells. Another strategy traced by nature is to fold the ciliary membrane, as found in ciliary photoreceptor cells (Eakin, 1968; Eakin, 1982). Initially it has been proposed that rhabdomeric photoreceptors are characteristic for protostomes, whereas ciliary photoreceptors are represented by deuterostomes (Eakin, 1968; Eakin, 1982). However it turned out that ciliary and rhabdomeric photoreceptor cells co-exist. This finding raised the question about the phylogenetic relationship of these two photo-receptor cell types. Some authors proposed that all photoreceptor cells can be traced back to a single precursor photoreceptor cell type present in Urbilateria. Based on this view, rhabdomeric and ciliary photoreceptor cells may have evolved multiple times independent from the urbilaterian precursor cell. An alternative perception is, in view of the widespread occurrence of both receptor types in bilaterian, that both ciliary and rhabdomeric photoreceptor cell types were already present in Urbilateria. New molecular data and the construction of phylogenetic trees for conserved proteins used in the phototransduction pathway and quenching, like opsin, opsin-coupled G-protein, arrestin and rhodopsin kinase, suggest that the two photoreceptor types represent distinct paralogs. These findings favour the view that the two photoreceptor cell types coexisted already in Urbilateria (Arendt, 2003). A plausible explanation is to argue that a single pre-bilaterian photoreceptor cell precursor diversified into two distinct types just at the outset of bilaterian evolution followed by subsequent gene duplication events and concomitant functional diversification (Fig 5.1.2).

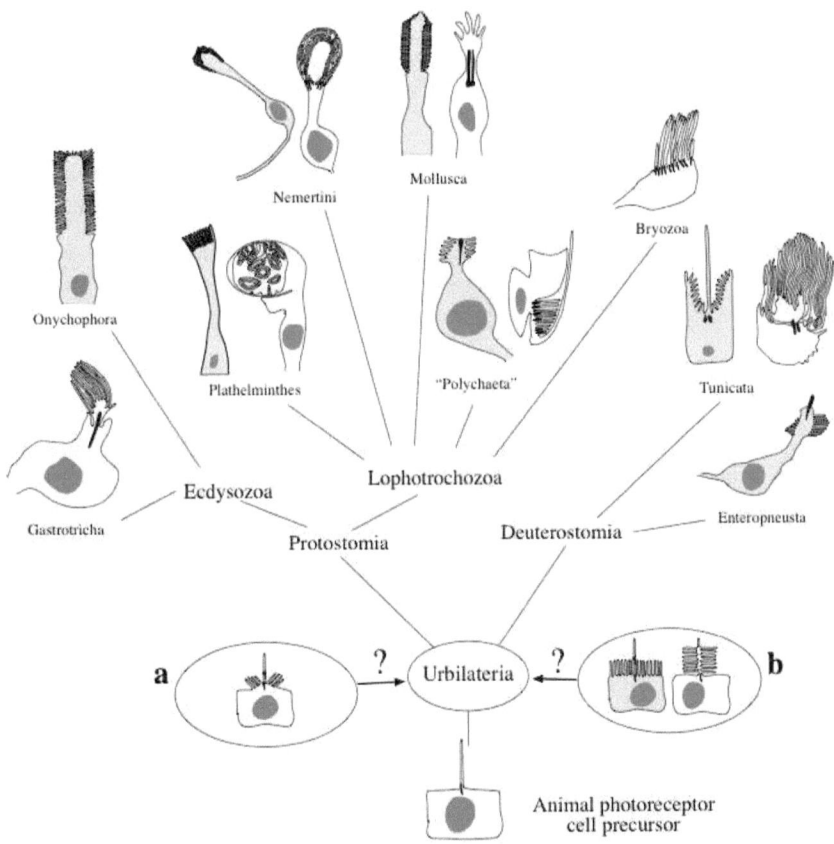

Figure 5.1.2 Two conflicting scenarios for the evolution of rhabdomeric and ciliary photoreceptor cell types. Alternative (a) considers the possibility that the ciliary and rhabdomeric photoreceptor cell types evolved multiple times independently, whereas alternative (b) suggests that both photoreceptor cell types emerged and coexisted already in Urbilateria. From (Arendt, 2003).

5.2 Usage of distinct phototransduction pathways in rhabdomeric versus ciliary photoreceptors

The first event in phototransduction is the absorption of a photon by the covalently bound retinal, most commonly 11-cis retinal, which initiates the isomerization to all-trans retinal (for review, see (Hargrave and McDowell, 1992). This is followed by a series of interactions between the retinal and the opsin protein, leading to a conformational change of opsin. The activated rhodopsin is now able to bind and activate a heterotrimeric G-protein. Upon activation, the α-subunit of the G-protein exchanges GDP for GTP and the α-subunit dissociates from the βγ-subunit (Hamm and Gilchrist, 1996). Up to this point the transduction events are shared by both the rhabdomeric as well as the ciliary photoreceptor. However the downstream events differ (Figure 5.2.1). In ciliary photoreceptors, the dissociated α-subunit of G-protein binds to cGMP-phosphodiesterase (PDE) and stimulates its hydrolytic activity by removing the inhibitory γ-subunit (Figure 5.2.1B). The activated PDE now catalyzes the hydrolysis of cGMP leading to a decrease of intracellular cGMP concentration (Lamb, 1996; Miki et al., 1973). The low level of cGMP finally leads to the closure of cGMP-gated cation channels on the cell membrane and results in hyperpolarization (for review, see (Arshavsky et al., 2002).

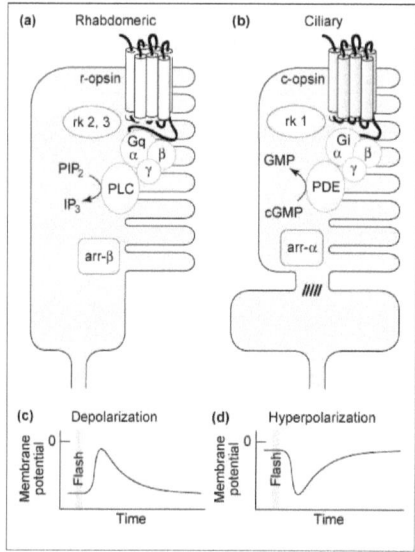

Figure 5.2.1 Usage of distinct phototransduction pathways in rhabdomeric versus ciliary photoreceptor cells leading to different physiological responses (see text for more information; from (Nilsson, 2004).

In contrast to ciliary photoreceptors, the α-subunit of the G-protein in rhabdomeric photoreceptors binds to the phosphoinositide-specific phospholipase C (PLCβ) (Figure 5.2.1A). The PLCβ then hydrolizes the membrane bound phospholipid, phosphatidyl inositol 4,5-bisphosphate (PIP_2) producing soluble inositol 1,4,5-trisphophate ($InsP_3$) and diacylglycerol (DAG). This results, by a yet unknown mechanism, in the activation of cation-permeable channels and to the depolarization of the membrane (Hardie, 2001a).

6. Genes involved in the genetic cascade of eye development

6.1 The Pax Gene family

6.1.1 Pax genes in general

The Pax family of transcription factors are characterized by a highly conserved 128 amino acid long DNA-binding domain, the Paired domain and a Paired type homeodomain. The first isolated gene containing a paired box was the segmentation gene *paired* from Drosophila (Bopp et al., 1986). In the meantime, many Pax genes were isolated from various metazoan species. The Pax genes can be grouped into four different classes depending on whether they have an octapeptide or not and whether they have a complete, partial or no homeodomain (Figure 6.1.1.1).

Data from crystallography indicate that the Paired domain consists of two independent subdomains, an amino-terminal PAI domain and a carboxy-terminal RED domain (Czerny et al., 1993; Xu et al., 1995)which recognizes a bipartite DNA site of about 17 nucleotides (Czerny et al., 1993; Epstein et al., 1994a). The two subdomains structurally resembles the helix-turn-helix (HTH) motif connected by a linker region. However, biochemical studies suggest that the isolated Paired domain does not adopt a fixed conformation unless it is incubated with DNA (Epstein et al., 1994a).

The PAI domain is generally more strongly conserved and seems to be dominant over the RED domain. Three amino acids (at position 42, 44 and 47) within the PAI domain are responsible for the different DNA-binding specificities between Pax2/5/8 and Pax6 (Czerny and Busslinger, 1995). Pax6 is specified by the amino acids IQN at these positions, whereas amino acids QRH define Pax2/5/8 specificity. However, recent evidence suggest that intramolecular interactions with distinct DNA-binding domains can modify the activity of Pax genes (Underhill and Gros, 1997). Moreover it has been shown that the Paired domain can also act as a protein-protein interaction domain (Plaza et al., 2001; Underhill and Gros, 1997). In addition to the Paired domain, the Pax genes have a second DNA binding domain, the 60 amino acid long Homeodomain. Most homeoproteins, including all Hox proteins bear a Gln at position 50. In contrast, Pax homeodomains are always characterized by a serine at the position 50, which is known to be crucial to determine the DNA binding specificity. Homeodomains were found to bind to palindromic TAAT-like target sequences, either as a homodimer or as heterodimers with other homeodomain containing transcription factors.

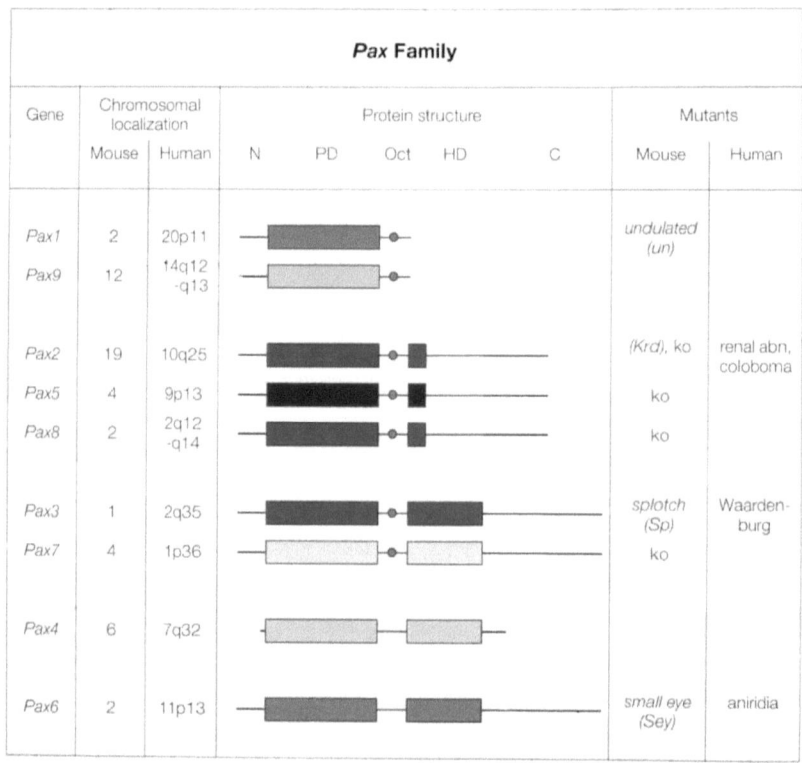

Figure 6.1.1.1 Schematic representation of the different Pax gene families in mouse and human. All Pax proteins contain a paired domain and, with the exception of the Pax4/6, an octapeptide. Pax2/5/8 have only a partial homeodomain and the homeodomain lacks completely in Pax1/9 (kindly provided by W. Gehring).

Alternatively splicing and alternative promoters are common mechanisms of Pax genes to modify the DNA binding characteristics. For example, alternative splicing within the Paired domain of Pax3, Pax6 and Pax8 alters the DNA-binding specificities (Epstein et al., 1994b; Kozmik et al., 1997; Vogan et al., 1996).

Pax genes (except Pax4 and Pax6) have an additional highly conserved eight amino acid domain, the octapeptide, located in the linker region between the Paired and the Homeodomain. The consensus amino acid sequence is HSIDGIL(G/S) for Pax3 and Pax7, YSI(N/S)G(I/L)LG for Pax2, Pax5 and Pax8, and HVS(S/T)(N/D)ILG for Pax1 and Pax9 (Noll, 1993). Deletion studies suggest that the octapeptide has inhibitory activity mediated by interaction of co-repressors, as for example the Groucho family (Eberhard et al., 2000; Lang et al., 2005).

Pax genes have been shown to play an important role in the development of various organs. Examples for organs where Pax genes seem to be crucial are the eye (Pax6 and 2), the skeleton (Pax1 and 9), the kidney (Pax2 and 8), B cells (Pax5), the thyroid (Pax8), the pancreas (Pax4 and 6), the central nervous system (Pax2, 3, 5, 6, 7, 8) and the skeletal muscle (Pax3 and 7).

6.1.2 Pax function in eye development

Pax genes are composed by different DNA binding domains, which may interact and cooperate. As a result they are capable to regulate a very broad spectrum of genes organized in networks or developmental programs. It has been proposed that the Paired domain and the Homeodomain, each able to regulate separate biological programs independently, might have been co-opted within a single Pax gene to regulate the development of the proto-type eye (Kozmik, 2005). There are two essential building blocks which have to be generated to build up a prototype eye: the dark pigment for shading and the photopigment to capture photons. Several lines of evidences suggests that there are specific roles for the Paired domain and Homeodomain in this process. It has been postulated that the Paired domain might be predominantly involved in pigmentation programs, regulation of crystallin expression and for general eye morphogenesis, whereas the Homeodomain is required for the expression of the photopigment gene *opsin*. In favour for such a model is for example the finding that the *Drosophila Pax2* homolog *sparkling*, which has only a partial homeodomain, is required for the development of pigment cells in the compound eye (Fu and Noll, 1997). Consistent with this finding, murine *Pax2* and *Pax6* are also expressed in the developing retinal pigment epithelium (Martinez-Morales et al., 2004). Pax6 and Pax2 were shown to bind and activate a retinal pigment epithelium specific *mitf* promoter element in vitro (Baumer et al., 2003). The microphthalmia-associated transcription factor, Mitf, has a conserved and fundamental function in the development of melanin producing cells and directly regulates melanogenic enzymes (Martinez-Morales et al., 2004). Mitf loss of function leads to a transdifferentiation of retinal pigmented epithelia into unpigmented retina, whereas overexpression induces pigmentation in the neuroretina. Consistent with this idea, Pax3 has found to be expressed in neural crest derived melanocytes (Martinez-Morales et al., 2004).

These findings may reflect an ancestral role of the Homeodomain in *opsin* regulation. In vertebrates however, Pax6 is not expressed in ciliary photoreceptors and is, thus, not used for activation of *opsin* genes. Interestingly, Pax6 is found to be expressed in the retinal ganglion cells of vertebrates, which are thought to be homologous to the ancestral rhabdomeric photoreceptor cell type (Arendt, 2003). In the cause of evolution other Homeodomain containing proteins were recruited to regulate *opsin* expression, such as Crx in vertebrates or otd in *Drosophila*.

It has been proposed that at the origin of modern Pax genes, a Paired domain containing protein (likely originated from a transposase) captured a Homeodomain through gene fusion, leading to a protein family able to bind complex cognate DNA sites (Breitling and Gerber, 2000). A Pax-B like gene mostly related to the Pax2/5/8 subfamily was isolated from the sponge *Ephydatia fluviatilus*, one of the most primitive representatives of the animal kingdom (Hoshiyama et al., 1998). The sponge Pax gene encodes for a degenerated but nevertheless well recognizable Homeodomain, suggesting that the Pax genes are of monophyletic origin which captured the Homeodomain very early in evolution and that Homeobox-free Pax genes evolved by losing the Homeodomain. It seems plausible therefore to postulate that the origin of Pax genes predates the origin of eyes and the nervous system. Based on studies of visual system development and the role of Pax genes in very basal animals, it seems that the Pax genes have a very ancient and fundamental role in eye development (Kozmik et al., 2003; Nordstrom et al., 2003; Piatigorsky and Kozmik, 2004; Sun et al., 2001).

6.2 *Pax6*

6.2.1 *Pax6* in general

Pax6 belongs to the Pax gene family of transcription factors and is highly conserved throughout the animal kingdom. It was first isolated from vertebrates, first mice and humans (Ton et al., 1991; Walther and Gruss, 1991), and shortly after it was cloned from zebrafish (Puschel et al., 1992). The human *Pax6* was isolated as a positional candidate for the ocular and neurodevelopmental disorder aniridia. For proper eye development one wild-type allele is not enough, hence heterozygous conditions lead to haploinsufficiency. Patients suffering from aniridia have ocular abnormalities (Prosser and van Heyningen, 1998; van Heyningen and Williamson, 2002) and moderate defects in the olfactory system and the brain (Ellison-Wright et al., 2004; Sisodiya et al., 2001). The clinical picture of aniridia includes iris hypoplasia, often combined with cataracts, corneal defects, foveal dysplasia, glaucoma, nystagmus and foveal and optic nerve hypoplasia (Figure 6.2.1.1) (Hittner, 1989; Nelson et al., 1984). About 80% of *Pax6* mutations in humans lead to typical aniridia phenotypes

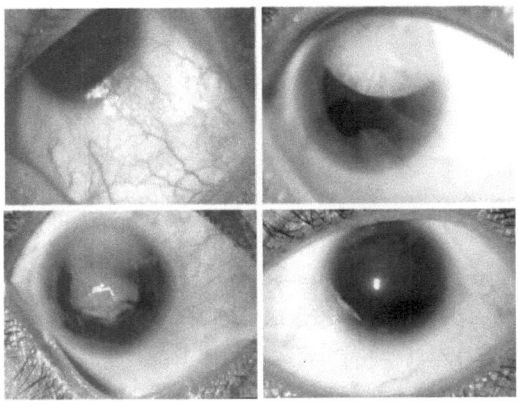

Figure 6.2.1.1 Heterozygous Pax6 mutations in humans result in sever to mild eye defects. From (Neethirajan et al., 2004).

(Prosser and van Heyningen, 1998), 10% of mutations involve regulatory mutations (Kleinjan et al., 2001; Lauderdale et al., 2000) and about half of the remaining 10% cases are missense mutations generating single amino acid substitutions which cause less severe phenotypes, e.g. foveal hypoplasia, Peter's anomaly, congenital cataracts and autosomal dominant keratitis (Prosser and van Heyningen, 1998; van Heyningen and Williamson, 2002). There are two reported cases of homozygosity, which led to anophthalmia, noseless phenotype and severe brain defects (Glaser et al., 1994). In mice, mutations in the *Pax6* gene results in small eyes, a phenotype very similar to that of human aniridia (Hill et al., 1991).

A *Pax6* homolog was also found in *Drosophila* (Quiring et al., 1994). The *Drosophila Pax6* homolog showed high sequence similarities in both the paired (94% identity) and the homeodomain (90% identity) to the vertebrate homolog. Very surprisingly it turned out that the *Drosophila Pax6* was the *eyeless (ey)* gene, known by a mutation affecting the eyes since 1915 (Hoge, 1915). This was a complete surprise because it was generally accepted that the camera eyes of vertebrates and the compound eyes of insects are non-homologues and have evolved independently. The finding that *Pax6* is not only highly conserved in sequence homology but also in its function led to the idea that *Pax6* might be a universal master control gene for eye development (Quiring et al., 1994). This hypothesis was further confirmed by targeted gene expression of *eyeless* in other imaginal discs than the eye disc using the Gal4 system (Halder et al., 1995). Ectopic eyes were induced on the legs, wings, halteres and the antennae of the fly. Another striking evidence for a strong functional conservation of *Pax6* was the finding that ectopic expression of murine *Pax6* can induce ectopic in the fly (Halder et al., 1995). Consistently, the reciprocal experiment, overexpression of *eyeless* or *twin of eyeless (toy)* in *Xenopus* embryos, leads to the development of vertebrate eye structures (Onuma et al., 2002).

6.2.2 Pax6 protein structure

Pax6 proteins have an N-terminally located 128 amino acid long Paired domain, a linker region of variable length, a 60 amino acid long Homeodomain and C-terminal proline-serine-threonine rich region (Figure 6.2.2.1A). The paired domain is a bipartite DNA recognition domain, separated in a N-terminal PAI and a C-terminal RED subdomain (Czerny et al., 1993; Xu et al., 1995). Both domains fold into a helix-turn-helix motif similar to the homeodomain and are separated by a short linker. The N-terminal PAI subdomain is build up by a short β-sheet, followed by a type II β-turn, three helices, which have a similar conformation as the homeodomain, and a C-terminal tail (Figure 6.2.2.1C). The N-terminal β-sheet interacts with the sugar phosphate backbone of the DNA. The following β-turn fits directly into the minor groove of the DNA and makes critical base contact. DNA binding specificity of Pax6 is determined by the amino acids at position 42, 44 and 47 of the Paired domain (Czerny and Busslinger, 1995). Isoleucine at position 42 and glutamate at position 44 are Pax6-specific, whereas the asparagine residue at position 47 is shared with Pax4. All other Pax proteins have a glutamate at position 42, an arginine at position 44 and a histidine at position 47. Three α-helices are following of which helix two and three fold into a helix-turn-helix motif (Figure 6.2.2.1C). Helix 2 makes contact to the DNA phosphate backbone, whereas recognition helix 3 fits into the major groove. The C-terminal tail of the PAI subdomain contacts the minor groove (Halder et al., 1995; Xu et al., 1995).

The C-terminal RED subdomain contains three helices folding into a helix-turn-helix motif. However, there is evidence that the RED domain of Pax6 is usually not involved in DNA binding. It is suggested that the PAI subdomain which provides the more important DNA contacts is sufficient for DNA binding (Cai et al., 1994; Chalepakis et al., 1991; Czerny et al., 1993; Treisman et al., 1991). However, there is a Pax6 splice variant known (Pax6 5a) which contains a 14 amino acid insertion in the PAI subdomain, which disrupts its DNA binding capability and enables the RED subdomain to make contact to a binding site other than the PAI consensus site (Epstein et al., 1994b). The crystal structure of a paired-type Homeodomain has been determined by X-ray christallography (Wilson et al., 1995). The homeodomain contains three helices folded into a globular domain which is organized into a flexible N-terminal arm, followed by Helix 1 that is separated from Helix 2 by a loop (Figure 6.2.2.1C). Helix 2 and Helix 3 form a helix-turn-helix motif, Helix 3 being the recognition helix. The N-terminal arm makes base-specific contacts with the minor groove, whereas Helix 3 makes contact with the mayor groove.

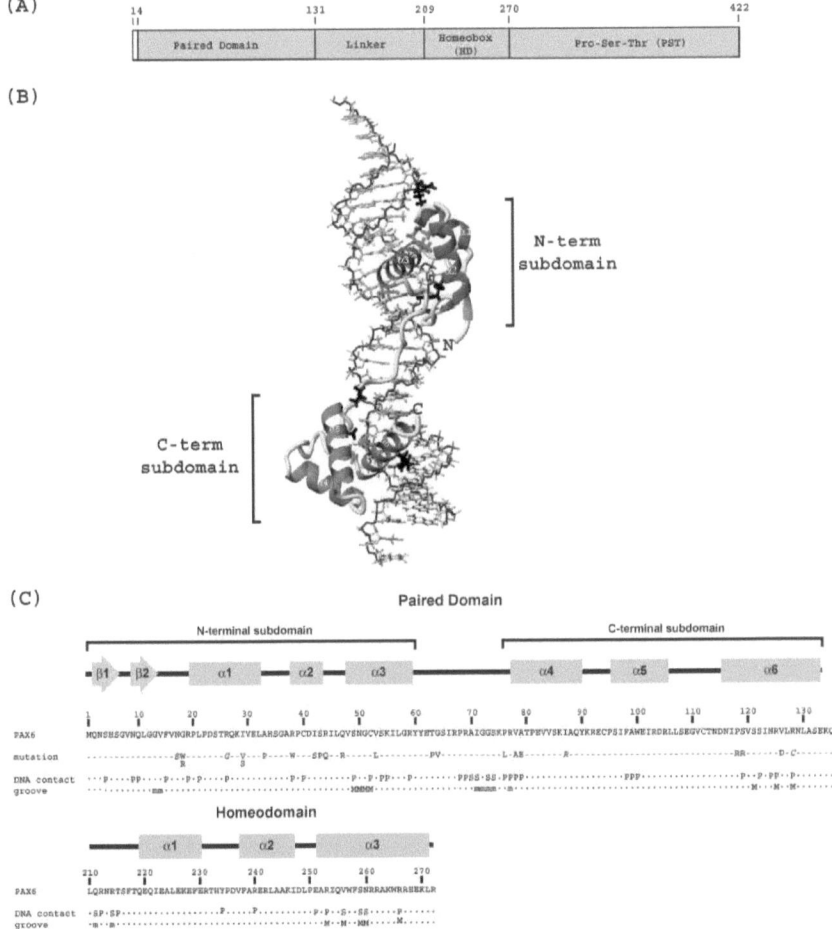

Figure 6.2.2.1 Pax6 protein structure. (A) Schematic representation of the human Pax6 protein. (B) The structural model of the Pax6 Paired domain in complex with a 26 bp DNA duplex. (C) The amino acid sequence of the Paired domain and Homeodomain. Protein interactions with DNA are indicated below the amino acid sequence; minor (m) groove, mayor (M) grove. Modified from (Tsonis and Fuentes, 2006).

6.2.3 *Pax6* expression in the vertebrate eye

Pax6 has been examined in various vertebrates as for example in the mouse (Ton et al., 1991; Ton et al., 1992; Walther and Gruss, 1991) and zebrafish (Krauss et al., 1991; Puschel et al., 1992). The following description is based on observations in the mouse, but expression is similar in other vertebrates. The first morphological indication of eye development starts with the formation of the optic pit. At this stage, *Pax6* is broadly expressed in the surface ectoderm including the optic pit. Subsequently, *Pax6* is expressed in the forming optic vesicle and the developing optic stalk. Later, as the optic cup starts to form, expression becomes weaker and subsequently vanishes in the optic stalk. In the optic cup, *Pax6* is expressed in a distal-proximal gradient. Expression of *Pax6* is first observed in both epithelial layers of the optic cup, but becomes restricted to the inner layer, the presumptive neuroretina, whereas expression in the outer layer, the presumptive retinal pigment epithelium, is only seen near the rim of the optic cup. This distalmost part of the retinal pigment epithelium will give rise to the future neuroectodermal components of the iris and the ciliary body. During retinogenesis, *Pax6* is found to be expressed in virtually all retinal progenitor cells. However, as more and more cells differentiate into diverse cell lineages, *Pax6* expression becomes restricted to distinct retinal cells (Marquardt et al., 2001). In the adult neuroretina expression is restricted to the amacrine neurons, the ganglion cell layer and the bipolar nerve layer.

In amphibians and fish the retina continues to grow throughout adult live to keep pace with the increasing body size. The new retinal cells are generated from a specialized proliferative zone in the peripheral rim of the neuroretina, the ciliary marginal zone (CMZ) (Wetts et al., 1989). The CMZ progenitor cells are multipotent and can give rise to all retinal cell types, including retinal pigmented epithelium cells (Wetts and Fraser, 1988). Similar to the retinal progenitor cells in the embryonic murine retina, *Pax6* is expressed in the distalmost stem cells of the CMZ (Perron et al., 1998).

During eye development, *Pax6* expression in the surface head ectoderm becomes progressively restricted to the developing lens placode, nasal placode and adjacent tissue. Subsequently, expression becomes further restricted, leading to separated expression in the lens and nasal placode. As the lens placode comes into close contact with the outgrowing optic vesicle, *Pax6* expression level is increased in the placode and becomes restricted to the proliferating lens epithelial cells. After the separation of the lens vesicle, *Pax6* remains expressed in the surface ectoderm destined to become the corneal epithelium (Grindley et al., 1995). In the lens, expression continues in differentiating epithelial fiber cells. In guinea pigs, *Pax6* expression has been shown to peak in the epithelium of the lens, is reduced in the equatorial region where cells differentiate into elongated lens cortical fibers and disappears in older layers (Richardson et al., 1995).

6.2.4 *Pax6* expression in the fly eye

Eyeless expression in *Drosophila* is first detected at the germ band stage. Expression is observed in a bilaterally symmetrical pattern in the brain, the anteriorly located primordia of the eye imaginal discs and in every segment of the ventral nervous system (Quiring et al., 1994). Later in development, expression is restricted to the brain region and the primordia of the eye disc. At the third larval stage, *eyeless* is expressed anterior to the morphogenetic furrow of the eye disc.

Expression of the second *Pax6* homolog of *Drosophila, twin of eyeless (toy)*, is first detected at the cellular blastoderm stage in the posterior cephalic region including the region of the presumptive optic lobe (Czerny et al., 1999). Subsequently, expression is observed in the dorsolateral head ectoderm which will give rise to the brain and the visual system, including the optic lobe, Bolwig's organ (larval eye) and the eye imaginal disc, which further develops into the adult compound eye and into three ocelli. After germband retraction, *toy* expression is found in the eye imaginal disc primordia. At the third instar larval stage, *toy* is expressed in the undifferentiated region of the eye disc which lies anterior to the morphogenetic furrow.

6.2.5 *Pax6* expression outside of the eye

Pax6 expression has been shown in many sites outside of the developing eyes. In mammals, *Pax6* is also expressed in the nasal placode, the pancreas, the gut, pituitary, brain and spinal cord at the early stages of embryonic development (Walther and Gruss, 1991).

During embryonic development, *Pax6* expression starts at the stage when the first somites are formed and the neural fold begins to close in the cervical region (Gérard, 1995; Grindley et al., 1995; Krauss et al., 1991; Puschel et al., 1992; Walther and Gruss, 1991). Expression is first detected in the prosencephalon (forebrain) and rhombencephalon (hindbrain), in the developing spinal cord and in a broad region of the head ectoderm. During forebrain development, *Pax6* expression is first detected in a broad domain in the neuroepithelium, including the region of the prospective optic vesicles, telencephalon and diencephalon (Grindley et al., 1995; Walther and Gruss, 1991). In the neural tube, *Pax6* expression is mainly restricted to mitotically active cells in the ventral ventricular zone and extends along the entire posteroanterior axis up to the rhombencephalic isthmus. As the head ectoderm starts to form the nasal and eye placode, *Pax6* expression becomes restricted to the placodes.

Pax6 is expressed in the nasal placode and stays expressed in the placodal epithelium during the formation of the nasal pit. Subsequently, expression is observed in the developing olfactory epithelium (Grindley et al., 1995; Walther and Gruss, 1991).

In early stages of mouse pancreas development, *Pax6* expression is first detected in the foregut/midgut endoderm from which the pancreatic bud develops. During further development, expression is maintained in a subset of cells of the dorsal and ventral pancreas. In newborn mice, expression becomes restricted to the islets of Langerhans. *Pax6* expression in the pancreas of newborn mice has been detected in cells of the endocrine tissue expressing insulin, glucagon, pancreatic polypeptide (PP), or somatostatin, corresponding to mature β-, α-, δ-, γ-endocrine cells, respectively (Hill et al., 1999; St-Onge et al., 1997). Furthermore, *Pax6* expression was also found in the enterocrine cells of the small and large intestine (Hill et al., 1999).

In *Drosophila*, the *Pax6* homolog *eyeless* is expressed throughout the development of the central nervous system and the eye (Quiring et al., 1994). During germ band extension, *ey* is expressed in neuroblasts of the head, which will develop into parts of the brain. In the ventral nerve cord, *eyeless* expression is restricted to three neuroblasts per hemisegment (Callaerts et al., 1997; Quiring et al., 1994). The ganglion mother cell and neuron cell progenies continue to express *ey* throughout the larval stages and in adult. The second *Pax6* gene of *Drosophila*, *toy*, starts to be expressed earlier than *ey*. *Toy* expression is first detected at the cellular blastoderm stage in the posterior procephalic region (Czerny et al., 1999). Until germband retraction, *toy* expression is confined to the head region anterior to the cephalic furrow. At this stage, *toy* expression is observed also in the ventral nerve cord, partially overlapping with the expression of *ey* (Czerny et al., 1999). Later in development *ey* and *toy* are also expressed in the mushroom bodies of the central nervous system (Callaerts et al., 2001; Kammermeier et al., 2001; Noveen et al., 2000), which are high-order brain centers for olfactory associative learning and elementary cognitive functions in the fly. *Ey* has been shown to have an important function in axonal differentiation of the mushroom bodies neurons (Callaerts et al., 2001; Kammermeier et al., 2001; Noveen et al., 2000).

6.2.6 The function of *eyeless/twin of eyeless* in *Drosophila* compound eye development

Beside *eyeless*, a second *Pax6* homolog, *twin of eyeless* was discovered in *Drosophila* which arose as a duplication event during insect evolution (Czerny et al., 1999). Both *Pax6* homologs were shown to be able to induce ectopic eyes when ectopically expressed in

Drosophila (Czerny et al., 1999; Halder et al., 1995). *Toy* was suggested to be upstream of *ey*. Indeed, further investigations revealed an eye-specific enhancer of *ey* which is directly regulated by *toy* (Hauck et al., 1999). Although *ey* and *toy* have partially redundant functions, loss-of-function mutants for both *ey* and *toy* suggested that *toy* is required for the formation of ocelli whereas *ey* is mainly involved in compound eye development (Kronhamn et al., 2002; Punzo et al., 2002). Consistent with this finding, *so* was found to contain an ocelli-specific enhancer which is regulated by *toy*, but not by *ey* (Punzo et al., 2002).

As just mentioned, *so* was found to be a direct target of *ey* and *toy*. The *so* eye-specific enhancer contains five binding sites for *ey* and *toy*. TOY can bind to all five binding sites, whereas EY recognizes only three of them.

So together with *eyes absent (eya)*, a protein phosphatase (Li et al., 2003; Tootle et al., 2003), are able to induce ectopic eyes when ectopically expressed (Pignoni et al., 1997), probably by feed back activation of *ey*. The existence of a positive feedback loop was confirmed by the finding of an eye-specific enhancer of the *eyeless* gene which is recognized by SO (Pauli et al., 2005). Moreover, it was shown that *so* regulates its own expression by an autoregulatory loop which is crucial for proper ocelli development.

Coexpression experiments suggested that EY, HH and DPP may function together as a complex to promote cell proliferation in the eye disc and to prevent the premature expression of the more downstream transcription factors *so*, *eya* and *dachshund*.

6.2.7 *Pax6* function in vertebrate eye development

During eye development, *Pax6* is expressed in different tissues and is required to regulate the expression of a broad range of genes. *Pax6* has been shown to play a crucial role in common genetic programs such as cell proliferation and cell differentiation.

Initially, *Pax6* is expressed in a broad domain of the anterior neural plate including the cells which will give rise to the optic vesicle. Interestingly, *Pax6* is dispensable for the formation of the optic vesicles, as $Pax6^{-/-}$ mice retain the capability to form the optic vesicle and also the establishment of the neuroretinal and the retinal pigmented epithelium domains is not perturbed (Grindley et al., 1995).

Although *Pax6* is not required for the formation of the optic vesicle, it is required for patterning the neuroretina. After optic cup formation, *Pax6* is downregulated in the optic stalk and the retinal pigment epithelium but remains expressed in the neuroretina (Macdonald and Wilson, 1997). Expression in the neuroretina is found in the proliferating retinal progenitor cells, while it gets downregulated in most cells upon differentiation. Investigations of the function of *Pax6* in the neuroretina showed that it is required to maintain the multipotency of retinal progenitor cells and for their normal proliferation. In *Pax6*-deficient retinal progenitor cells, differentiation is reduced and exclusively amacrine interneurons are produced. Importantly, *Pax6* was shown to be required for the expression of several proneural genes, including the retinal basic Helix-Loop-Helix genes *Ngn2*, *Mash1* and *Math5* (Marquardt et al., 2001). Since retinogenesis has been shown to be normal in mice where the lens is eliminated by deleting lens-specific *Pax6* expression using the Cre/*loxP* system, *Pax6* is thought to act autonomously in the neuroretina (Ashery-Padan et al., 2000).

Several findings suggest that *Pax6* plays an important function during early stages of lens induction. In chimeric embryos of wild-type and *Pax6⁻/Pax6⁻* cells, for example, *Pax6⁻/Pax6⁻* cells are excluded from the surface ectoderm (Collinson et al., 2000). Furthermore, the lens-specific marker fails to be expressed in *Pax6⁻/Pax6⁻* mutant mouse embryos (Furuta and Hogan, 1998; Wawersik et al., 1999). Moreover tissue recombinations between the optic vesicle and the surface ectoderm of *Pax6⁻/Pax6⁻* and wildtype rat embryos suggest that *Pax6* is not essential for the inductive activity of the optic vesicle but rather has a cell autonomous function in the surface ectoderm (Fujiwara et al., 1994). By deleting *Pax6* exclusively from the surface ectoderm after the lens bias stage using the Cre/*loxP* system, all lens structures failed to form (Ashery-Padan et al., 2000). Interestingly, at this stage *Pax6* is not required to maintain the expression of the lens marker *Sox2*. Based on these results, it is suggested that *Pax6* function is essential in two successive stages of lens induction. Initially, *Pax6* is required to maintain the lens forming fate of the surface ectoderm by activating *Sox2* in the ectoderm. In the second stage, *Pax6* activity is necessary to initiate lens differentiation by controlling the expression of other lens specific regulatory genes such as *Six3*, *Prox1* and *Eya* (Ashery-Padan et al., 2000). *Six3*, for example, has shown to be able to induce ectopic lenses in fish (Oliver et al., 1996). Although at this stage *Pax6* is not required to maintain *Sox2* expression, *Sox2* alone can not support lens formation in the absence of *Pax6*. This is consistent with the finding that Pax6 binds cooperatively with Sox2 to the δ crystallin enhancer to mediate δ crystallin expression (Kamachi et al., 1999; Kamachi et al., 2001. Blanco et al., 2005).

6.3 The Six family genes

The first Six family gene isolated was *sine oculis* of *Drosophila*, which is known to have an important function for compound eye morphogenesis (Cheyette et al., 1994; Serikaku

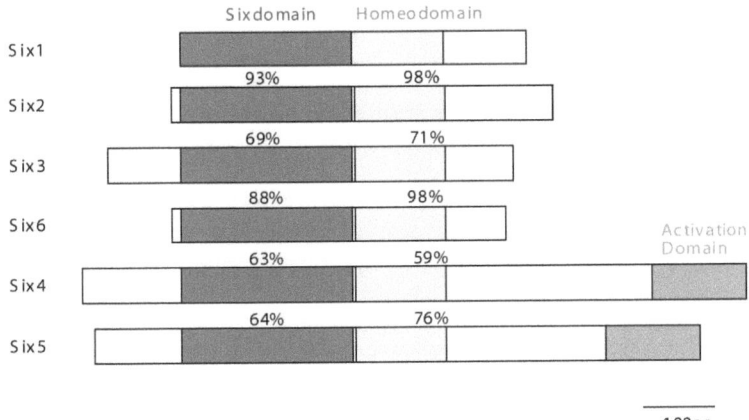

Figure 6.3.1.1 Structure of the Six family proteins. Schematic representation of the mouse Six proteins showing their characteristic structural features. The Six domain (blue) and the Six-type homeodomain (yellow). The percentage of amino acid identity between the Six domains and between the Homeodomains are indicated (Courtesy of D. Graziussi).

and O'Tousa, 1994). Subsequently, Six family genes were identified in many other species, such as humans (Boucher et al., 1996; Boucher et al., 1995), mouse (Kawakami et al., 1996; Oliver et al., 1995b), chicken (Bovolenta et al., 1998), frog (Seo et al., 1999), fish (Seo et al., 1999; Seo et al., 1998) nematode (Seo et al., 1999) and planaria (Pineda et al., 2000).

A common feature of the Six proteins is the highly conserved Six-domain of 110 to 115 amino acids. Beside that, there is another conserved region, the 60 amino acid long Six-type homeodomain. Characteristic for this homeodomain type is the lack of two highly conserved amino acids generally found in most homeodomains; an arginine at position 5, which is known to be important for the recognition of the DNA binding core sequence TAAT, and a glutamine at position 12 in helix 1 (Serikaku and O'Tousa, 1994).

In mice, six members of the protein family have been identified (Figure 6.3.1.). Based on sequence homologies in the Six domain and the Homeodomain, they can be grouped into three subclasses: Six1/2, Six3/6 and Six4/5 (Seo et al., 1999). In *Drosophila*, three members have been identified, *sine oculis*, *optix* and *Dsix4* (Seo et al., 1999). Further, four members have been isolated from *Caenorhabditis elegans* (Dozier et al., 2001) and at least one of each subclass in basal metazoans (Bebenek et al., 2004; Stierwald et al., 2004). The *Drosophila sine oculis*, which belongs to the Six1/2 subfamily, is required for the development of the entire visual system, including the Bolwig's organ (the larval photoreceptors), the three ocelli and the adult compound eye. *Sine oculis* has been shown to belong to the eye regulatory network together with *eyeless*, *twin of eyeless* (*toy*), *eyes absent* and *dachshund* and to be a direct target of EY and TOY (Niimi et al., 1999; Punzo et al., 2002). *Optix*, the *Drosophila* Six3/6 homolog, is expressed in the eye imaginal disc and has the capability to induce ectopic eyes in an *eyeless*-independent mechanism (Seimiya and Gehring, 2000; Toy et al., 1998). The third member, *Dsix4*, was shown to be important for myoblast fusion (Kirby et al., 2001).

In vertebrates, *Six3* and *Six6* are expressed in the developing brain and eye (Kobayashi et al., 1998; Loosli et al., 1999; Zuber et al., 1999) and at least *Six3* is suggested to be under direct control of *Pax6*.

Six5, has been shown to be critical for cataractogenesis and spermatogenesis (Sarkar et al., 2000; Sarkar et al., 2004). *Six1* and *Six2* are involved in a broad array of developmental processes. However, no function has been found for eye development. They are expressed in the head mesoderm and are involved in mesodermal patterning in the mouse and limb tendon development (Oliver et al., 1995a). *Six1* was found to be expressed in the otic and olfactory placode and is crucial for proper development of the nose and the auditory system (Laclef et al., 2003b; Zheng et al., 2003). Moreover, *Six1* is also suggested to play an important role in myogenesis, since mice $Six^{-/-}$ embryos show selective loss of muscles, including distal forelimb and hindlimb muscles and abdominal muscles (Laclef et al., 2003a). Consistent with a putative role in myogenesis, *Six1* and *Six4* have been shown to be required for the expression of *Pax3* and myogenic regulatory factors (Grifone et al., 2005).

7. The opsin gene family

The opsins represent a large family of G-protein coupled receptors (GPCR), also know as 7-transmembrane domain receptors, found in organisms ranging from archaea to humans (Spudich et al., 2000). Common characteristics of opsin apoproteins are the seven transmembrane α-helices and the covalently linked chromophore (retinal) which is attached to the ε-amino group of a lysine residue in the seventh helix through a Schiff base (Figure 7.1.1). Their most prominent and most extensively studied function is their role in vision. Beside that, opsins have very diverse functions, such as light-driven ion pumps, which mediates phototaxis, or more specialized function such as the processing of retinaldehyde isomers. Based on their various functions and sequence comparison, they can be divided into two

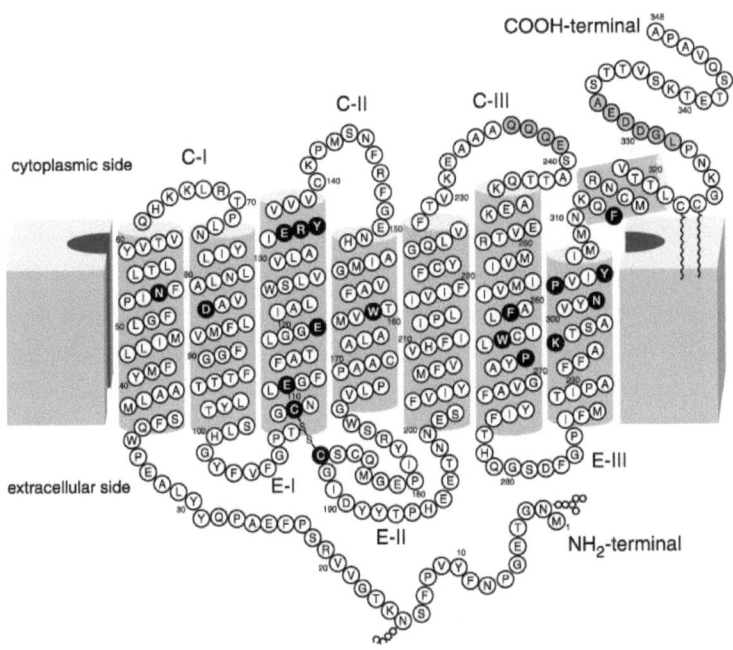

Figure 7.1.1 Two-dimensional model of bovine rhodopsin. Some of the key residues are indicated by filled circles. Residues in grey circles are not modeled in the current structure. From (Palczewski et al., 2000).

clearly distinct families, type 1 and type 2 rhodopsins (for review, see (Spudich et al., 2000). The archeal type 1 rhodopsin functions as light-driven ion transporters (bacteriorhodopsin and halorhodopsin), as receptors involved in phototaxis and many yet unknown functions (as for example in fungi, (Brown, 2004)). Type 2 rhodopsins include the photosensitive receptor proteins in animal eyes, as well as receptor proteins in the pineal gland, hypothalamus and other tissues of lower metazoans. All type 2 photoreceptors isolated so far are from higher eukaryotes. Although there is almost no or only little sequence homology between type 1 and type 2 rhodopsins, the three-dimensional protein architectures are quite similar. They both share the 7-transmembrane architecture and in all known cases a retinal chromophore is linked to a lysine residue in the 7th helix by a Schiff base. The fact that microbial-type rhodopsins have very little sequence homology to animal-type rhodopsins suggests an independent origin. It might be possible that animal rhodopsin evolved from other rhodopsin-related proteins which were originally unrelated to light-sensing receptors as for example chemoreceptors. Evidence supporting this idea is the finding that rhodopsin and the AMP receptor of *Dictyostelium*, which functions as an chemoattractant receptor, show significant sequence homology (Gehring, 2004; Klein et al., 1988). Further support for this idea comes from the fact that *Pax6* was shown to be critical not only for the development of the eye but also for the nose. These findings raise the possibility that the visual system might have evolved from a chemoreceptive system (Gaines and Carlson, 1995; Gehring, 2004).

Phylogenetic analysis of visual opsins employed by rhabdomeric and ciliary photoreceptors suggests that they represent two distinct orthologous opsin groups, "rhabdomeric opsins" and "ciliary opsins", respectively (Arendt et al., 2004). Interestingly, vertebrate ciliary opsins show higher homology to retinochromes than to rhabdomeric opsins. Reciprocally, invertebrate rhabdomeric opsins show higher homology to vertebrate melanopsin than to vertebrate ciliary opsins (Provencio et al., 1998; Provencio et al., 2000). In terms of evolution this is interesting because it suggests that they may trace back to distinct genes in Urbilateria. A plausible explanation would be to propose a single, pre-bilaterian photoreceptor cell which was using an ancestral opsin for light detection. During the course of evolution, the pre-bilaterian opsin duplicated to yield two opsin paralogs, the rhabdomeric and the ciliary opsin. Subsequently, diversification gave rise to rhabdomeric and ciliary sister cell types (Arendt et al., 2004).

The finding that opsins are also expressed in photoreceptive, non-visual tissues such as the pineal gland or the skin, lead to the informal classification of opsins as either "visual" or "non-visual" (Kawamura and Yokoyama, 1998; Kojima and Fukada, 1999; Van Gelder, 2004).

Interestingly, some recently identified opsins in vertebrates, which are not directly involved in image forming functions, were found to reside in the retina, although not in the rod or cone photoreceptors (Blackshaw and Snyder, 1999; Provencio et al., 2000; Sun et al., 1997; Tarttelin et al., 2003). This non-visual opsins found in the retina may play secondary roles in vision as for example to generate appropriate retinoid isomers for the visual opsins (Hao et al., 2000).

An important function of non-visual opsins is their involvement in photoentrainment. Almost all animals have an "internal clock" to keep pace to the 24h day and hence to the varying demands of day and night. To link the internal clock to the environmental circadian rhythm, photoreception, respectively opsins are used as an interface.

In mammals, the master circadian pacemaker is the suprachiasmatic nuclei (SCN) (Ralph et al., 1990). The light information for the SCN originates in the retina and runs through the retinohypothalamic tract (RHT), formed from about 1% of total retinal ganglion cells (Moore and Lenn, 1972). In mammals, photosensitive receptors involved in photoentrainment seems to be restricted to ocular photoreceptors since bilateral removal of the eyes have been shown to abolish photoentrainment (Nelson and Zucker, 1981). Several experiments showed that melanopsin expression is absent in rods and cones but are restricted to the small subsets of RGC projecting to the SCN. These findings suggested that the melanopsin expressing RGC might serve as photoreceptors to provide photic information to the circadian pacemaking SCN (Provencio et al., 1998; Provencio et al., 2000; Sollars et al., 2003).

In contrast to mammals, non-mammalian vertebrates have additional photoreceptor organs developing from the forebrain (Korf, 1994). All non-mammalian vertebrates have a pineal organ containing photoreceptors and deep brain photoreceptors, located at several sites in the brain, which are important to regulate circadian physiology. Moreover, in teleost fishes, a novel opsin family was recently found which are expressed in a wide array of tissues and which are suggested to be involved in the photic regulation of peripheral clocks (Moutsaki et al., 2003).

II. Material and Methods

1. Molecular methods

Standart molecular methods like DNA digestion, alkaline phosphate treatment, phenol-chloroform extraction of DNA, ligation, miniprep, agarose gel electrophoresis etc, were performed according to (Sambrook and Russel, 2001) and will not be further described. Only additional information about some protocols less commonly used are given in this section.

2. Collection of the animals

Arca noae were collected by Scuba divers from the Laboratoire Arago, Banyuls-sur-mer (South of France; Mediterranean Sea), whereas *Pecten maximus* were obtained from Roscoff (Northern France; Atlantic coast). The animals were transported to Basel and kept in a seawater aquarium.

Pecten maximus larvae were collected during one week at IFREMER in Brest (France). The larvae were directly fixed in 4% Paraformaldehyde and transported to Basel. There, Paraformaldehyde was substituted by Methanol and the larvae were kept at -20°C until use.

3. Preparation of genomic DNA

Gonads were dissected and 3g of the tissue was homogenized in 15ml HB buffer using a Polytron homogenizer. Equal volume of 1:1 Phenol/Chloroform was added and mixed by vortexing until well mixed and eventually shaken for 30 minutes on a flash shaker.
The extraction mix was centrifuged at 18,000g for 10 minutes at 20°C and the aqueous (top) phase was transferred eventually to a new tube. The extraction step was repeated twice. Two volumes of 100% ethanol was added to the aqueous phase transferred to a new tube, mixed and centrifuged at 18,000g at 20°C for 10 minutes. Eventually the pellet was washed with 70% ethanol and resuspended in TE.

4. Isolation of mRNA and cDNA synthesis

Messenger RNA was extracted using the Dynabead® mRNA DIRECT KIT™ from DYNAL® Biotech. For cDNA synthesis the SuperScript™ III First-Strand Synthesis System for RT-PCR from Invitrogen was used and carried out as described by the manufacturer.

5. Cryosections

Open the mussel with strong dissection scissors.

Carefully dissect the mantle edge containing the eyes lined up in a row of black dots (*Arca*) or individual eyes (*Pecten*) with small scissors,.

Tissue was fixed in 4% Paraformaldehyde

Embed in O.C.T.-compound Tissue-Tek (Miles Laboratories) and freeze in liquid nitrogen.
Mount onto metal block-holder and adjust temperature for sectioning at -20°C.
The thickness of the section is 12µm.
Collect the sections on Superfrost slide and immediately flatten on a 50°C hot plate for 1-2 min.
Let the slides air-dry for 1 hour to over night.

Postfix the slides	30min	4% Paraformaldehyde in PBS
Wash:	2x 10min	PBS, 01% Tween-20
Dehydrate	5min each	Ethanol series 30%, 70%, 95%, 100% EtOH/PBS

6. In situ hybridization protocol

1. Substitute MetOH with water using 75%, 50%, 25% MetOH (each 30')
2. Wash with DEPC water 10' @ RT
3. Wash with PBS-Tween 10' @ RT

4. Postfix with 4% PFA for 30'
5. Wash twice with PBS-Tween for 15'
6. Prehybridization for 1h @ 55°C (put on rotator)
7. Substitute Prehybridization solution with Hybridization solution (50ng Dig-labeled probe/ml) for 2h @ 55°C (rotator)
8. Substitute solution with 65°C prewarmed HB + probe (overnight @ 55°C, rotator)
9. Wash 3x with Wash solution for 30' @ 55°C (rotator)
10. Wash once with Wash solution for 30' @ 60-65°C (rotator)
11. Wash 4x with SolutionIII @ 60-65°C (2x 10', 2x 15'; rotator)
12. Add Solution III + BufferI (1:1) ; 20' @ RT
13. Wash 3x with BufferI @RT
14. Block with BufferII for 1h @ RT
15. 1/2000 anti-DIG/BufferII for 3h @ RT
16. Wash 3x with BufferI for 60'
17. Wash with TMN
18. BCIP-NBT/TMN
19. Stop reaction in TE or T-PBS

Solutions

Prehybridization solution:

50% Formamide, 5xSSC, 100μg/ml tRNA, 100μg/ml Heparin, 0.1% Tween20, 10mM DTT

50ml Prehybridization solution:

Formamide	: 25.0ml
20xSSC	: 12.5ml
1M DDT	: 0.5ml
Heparin (50mg/ml)	: 0.1ml
yeast tRNA (10mg/ml):	0.5ml
QH$_2$O	: 11.4ml

Hybridization solution:

50% Formamide, 5xSSC, 100µg/ml tRNA, 100µg/ml Heparin, 0.1% Tween20, 10mM DTT, 10% Dextran

50ml Hybridization solution:

Formamide	: 25.0ml
20xSSC	: 12.5ml
1M DDT	: 0.5ml
Heparin (50mg/ml)	: 0.1ml
yeast tRNA (10mg/ml):	0.5ml

Dissolve 5mg Dextran in ~8ml of Water @ 55°C and add to the other components. Fill up to a total volume of 50ml with QH_2O.

Wash solution:

50% Formamide, 5xSSC, 0.1% Tween20

Solution III:

50% Formamide, 2xSSC, 0.1% Tween20

Buffer I

0.1M Maleic acid, 0.15M NaCl, 0.1% TritonX

Buffer II

1% Blocking solution in Buffer I

TMNT:

100mM Tris-HCl pH9.5, 100mM NaCl, 50mM $MgCl_2$, 0.1% Tween20

50ml TMNT:

5M NaCl	: 1.0ml
1M MgCl	: 2.5ml
Tris-HCl 9.5	: 2.5ml
10% Tween20	: 0.5ml

BCIP-NBT/TMNT (10ml)

TMNT:	10.0ml
NBT (75mg/ml 70% DMF):	18.0µl
BCIP (50mg/ml DMF):	35.0µl

7. PCR Protocols

Amplification of Pax6 fragments by degenerated primers

initial denaturation	2min	at	94°C	
denaturation	30sec	at	94°C	
annealing	30sec	at	50°C	40x
elongation	1min	at	72°C	
final elongation	5min	at	72°C	
stop	∞	at	4°C	

Amplification of Six fragments by degenerated primers

initial denaturation	2min	at	94°C	
denaturation	30sec	at	94°C	
annealing	30sec	at	42°C	40x
elongation	1min	at	72°C	
final elongation	5min	at	72°C	
stop	∞	at	4°C	

Amplification of Opsin fragments by degenerated primers

initial denaturation	2min	at	94°C	
denaturation	30sec	at	94°C	
annealing	30sec	at	52°C	40x
elongation	1min	at	72°C	
final elongation	5min	at	72°C	
stop	∞	at	4°C	

7.1 Degenerated Primers

Pax6 degenerated primers

PrdX5 (YYETG) forward: 5'CAGCTCGAGNTAYTAYGARACNGG3'
Prd 33 (WEIRD) reverse: 5'GTATCTAGAGTCNCGDATYTCCCA3'

Six degenerated primers

Six-FP (PR(T/S/C)IW) forward: 5'TTYCCIYTICCIMRIWSIATITGGGA3'
Six-RP: (TQV(G/S)NWF) reverse: 5'TTYTTRAACCARTTISIIACYTGIGT3'
SixDomainF (QVACVC) forward: 5'CARGKBGCBWGYGTBTGYGA3'
SixHD-AS (NWFKNRRQR) reverse: 5'CKKCKGTTYTTRAACCARTTGSWVAC3'

Opsin degenerated primers

opsin-1s (forward): 5'TGGGCIIIIIIICCIITIITNGGNTGG3'
opsin-2s (forward): 5'GCCTTYITIITIRCITGGWCNCCNTA3'
opsin-1a (reverse): 5'TTGGACAIIMCRTAIAYIAINGGRTT3'
opsin-2a (reverse): 5'AACGCTAIIAIIGMRTAIGGNGTCCA3'

8 RACE PCR

Race PCR was performed using the 3' RACE system for Rapid Amplification of cDNA Ends Version E/ 5' RACE system for Rapid Amplification of cDNA Ends Version 2.0 from Invitrogen (Catalog No 18373-019 and Catalog No 18374-058). Experimental procedures were carried out as described in the manufacturer's protocol.

8.1 RACE primers

AnPax6 **3' RACE**
AnP63'race1 (forward): 5'CGACCACGTGCAATCGGCGGTAGC3'
Anp63'race2 (nested forward): 5'AGCAAGCCAAGAGTAGCCACAAATGAT3'

AnPax6 **5' RACE**
AnP65'race1 (reverse): 5'GCTACCGCCGATTGCACGTGGTCG3'
AnP65'race2 (nested reverse): 5'ATCATTTGTGGCTACTCTTGGCTTGCT3'

PmaPax6 **3' RACE**
PmaP63'race1 (forward): 5'CGAGCAATAGGCGGTAGTAAGCCCAGAG3'
PmaP63'race2 (nested forward): 5'GCCCAATACAAGAGGGAGTGTCCGTCA3'

PmaPax6 **5' RACE**
PmaP65'race1 (reverse): 5'GTATTGGGCTATTTTGCCTACGACGTC3'
PmaP65'race2 (nested reverse): 5'CACTCTGGGCTTACTACCGCCTATTGCTCGAG3'

AnSix1/2 **3' RACE**
AnSix3'race1 (forward): 5'CCGAGGACCATTTGGGATGGGG3'
AnSix3'race2 (nested forward): 5'GCACATAATCCTTATCCTTCCCCG3'
AnsSix1/2 **5'RACE**
AnSix5'race1 (reverse): 5'GGCATCCAAACATTCATAGCAC3'
AnSix5'race2 (nested reverse): 5'GGTGATTGTTTGGGACTCATCGGC3'
PmaSix1/2 **3'RACE**
PmaSix3'race1 (forward): 5'TGGGACGGTGAGGAGACCAG3'
PmaSix3'race (nested forward): 5'GACTGGTATTCCCACAATCCCTAC3'
PmaSix1/2 **5'RACE**
PmaSix5'race1 (reverse): 5'ATGAAAAGCCACTACGGCCTTAGC3'
PmaSix5'race2 (nested reverse): 5'GTTCACAAGCTGGCAATGACC3'
AnOpsinX **3'RACE**
AnOps3'race1 (forward): 5'CCAACTTGAAGAACAGCCACACCCTGC3'
AnOps3'race2 (forward): 5'GAGGCGATATCACAAGTCGTGGTATC3'
AnOpsinX **5'RACE**
AnOps5'race1 (reverse): 5'GTCTGCTAGAGACTGGCCTGTCC3'
AnOps5'race2 (nested reverse): 5'CGTACACGATGTACCATATGCTTCTATGG3'
PmaGqOpsin **3'RACE**
PmaGqops3'race1 (forward): 5'GCAAGGGAAATGGGCAGCATGG3'
PmaGqops3'race2 (nested forward): 5'GTCATGGAGTCCCTACGCTAC3'
PmaGqOpsin **5'RACE**
PmaGqops5'race1 (reverse): 5'GTAGCGTAGGGAGTCCATGAC3'
PmaGqops5'race2 (nested reverse): 5'CCATGCTGGCCATTTCCCTTGC3'
PmaOpsinX **3'RACE**
PmaOps3'race1 (forward): 5'TGTTCAATAGACTGGACTTC3'
PmaOps3'race2 (nested forward): 5'CTTCTCCTGATAGGCCTGGTG3'
PmaOpsinX **5'RACE**
PmaOps5'race1 (reverse): 5'CACCAGGCCTATCAGGAGAAG3'
PmaOps5'race2 (nested reverse): 5'GAAGTCCAGTCTATTGAACA3'

9. Real-time quantitative PCR

Messenger RNA was extracted from different tissues using the Dynabeads® mRNA DIRECT KIT™ (DYNAL® Biotech) and reverse transcribed using the SuperScript™ III First-Strand Synthesis System (Invitrogen). From the obtained cDNA, 1μl was used as a template for quantitative PCR amplification. Real-time PCR was performed by using a Light Cycler (Roche) and QuantiTect™ SYBR®Green PCR kit (Qiagen) under the following conditions: 15 minutes at 95°C, 15 seconds at 94°C, 20 seconds at 56°C, 10 seconds at 72°C over 50 cycles. The fluorescence of the amplified products was analyzed during 5 seconds at 76°C after the elongation step. For each sample point, a melting curve was obtained at the end of the PCR amplification to verify the specificity of the amplicon. In each experiment and for each gene, a standard curve generated by four dilutions from each cDNA sample was included in the PCR amplification to determine the expression value for each sample. As a housekeeping gene for normalization of the data *Elongation factor 1 alpha* (Ef1α) was used. The following specific primers used for real-time PCR were designed using a primer software at (www.genscript.com/ssl-bin/app/primer).

9.1. Primers for real-time PCR

AnPax6:
forward: 5'CACCATATCCAACCCATCAA3'
reverse: 5'GCGCATTTGTTCATCAGACT3'
PmaPax6
forward: 5'GTCAACCAGTTGGGAGGAGT3'
reverse: 5'CCATTGGAAACCTGGAGAAT3'
AnSix1/2
forward: 5'ATATTTAAAGCGGGCTGTGG3'
reverse: 5'AGGGTGATTGTTTGGGACTC
PmaSix1/2
forward: 5'CACAATCATCCCAAACTCCA3'
reverse: 5'CACCGTCCCAAATAGTCCTT3'

AnOpsinX
forward: 5'AATATTGGGCGTAATTTGGG3'
reverse: 5'GGCCTGTCCAGTCAATCGTA3'
PmaOpsinX
forward: 5'CTCCTGATAGGCCTGGTGTT3'
reverse: 5'ACTTTCACGTCCCTCTTGCT3'
PmaGqOpsin
forward: 5'GCATGGCTGACAAACTCAAC3'
reverse: 5'ATCTCCAAACTGGGCCATTA3'
Anefla
forward: 5'TCGGGTACTGGTGAATTTGA3'
reverse: 5'GGCCTCAGAGTATGGTGGTT3'
Pmaefla
forward: 5'GGACAGTACAGAGCCACCCT3'
reverse: 5'CTCGATCATGTTGTCACCGT3'

10. Targeted expression of AnPax6 and PmaPax6 in Drosophila

Ectopic expression of *AnPax6* and *PmaPax6* in larval imaginal discs of D. melanogaster was performed using the Gal4 system (Brand and Perrimon, 1993). Full-length *AnPax6* and *PmaPax6* cDNAs were inserted as *BglII-NotI* and *Asp718-NotI* fragments, respectively, into the pUAST vector. For each construct, transformant flies from eight independent transgenic lines were crossed to the dpp^{blink}-Gal4; UAS-Gal4 driver line.

11. Scanning electron microscopy

For scanning electron microscopy, freshly hatched flies were narcotized and immersed in 3% Glutaraldhehyde for 5 hours at room temperature. Subsequently, an ethanol dehydration series was carried out. After critical point drying, they were mounted and coated with gold.

III. Results (*Arca*)

1. Ultrastructure of the Arca noae compound eye

Individuals of *Arca noae* possess at least 100 or more eyes on each of their halves. The eyes are situated on the outermost fold of the mantle edge (Figure 1.1D). This is in contrast to most other eye-bearing bivalves, which usually carry their eyes on the middle fold of the triply folded mantle edge (Waller, 1980). The eyes are mainly concentrated in the posterior and anterior part of the mantle edge with the highest number anteriorly. The number

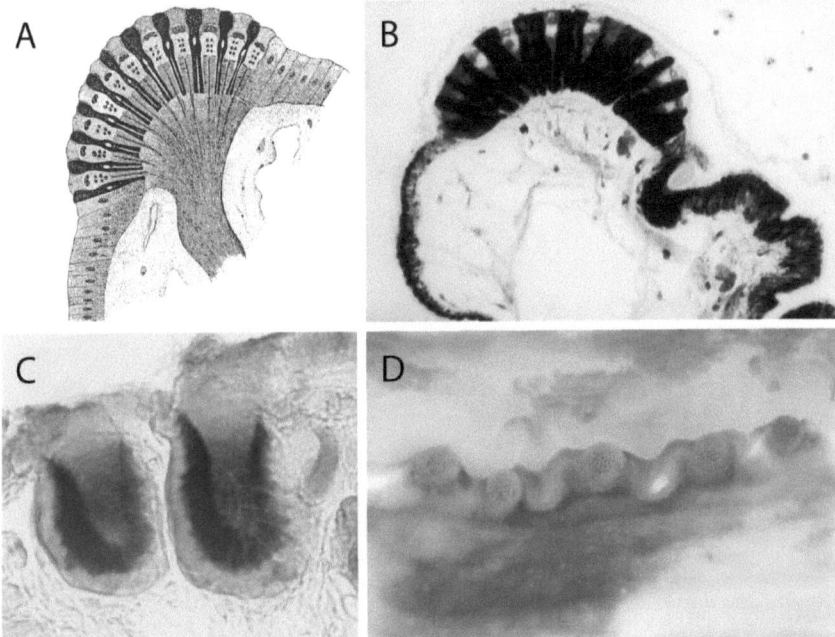

Figure 1.1 The eyes of *Arca noae*. A) Schematic drawing of an *Arca* compound eye (after Küpfer). B) Cross-section through an *Arca* compound eye (Courtesy of Heinz Streble). C) Cross-section through two pit eyes. D) The mantle edge with a row of compound eyes.

of eyes gradually decrease towards more ventral regions, with no or only a few eyes at the most ventral part of the mantle edge. Toward the posterior end, the number of eyes increases again. The compound eyes are very variable in size. Larger eyes are generally found in the most anterior regions, however also smaller eyes were found to be intermingled between larger ones.

The compound eyes appear heavily pigmented, disrupted by a regular array of hollow tubes surrounded by pigment cells representing the ommatidial units (Figure 1.1B). This arrangement of densely packed ommatidia gives the whole eye an appearance more resembling a sponge than an eye.

Additionally, numerous small pit eyes located around the compound eyes were observed (Figure 1.1C).

In contrast to the compound eyes of the ark clam *Barbatia cancellaria*, which were shown to have a concave depression in the centre (Nilsson, 1994), the eyes of *Arca noae* are of almost perfect convex shape.

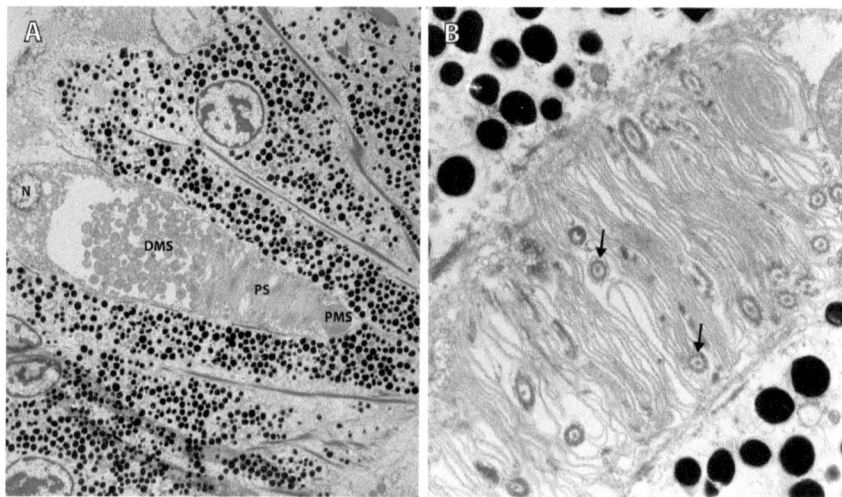

Figure 1.2 Electron micrographs of the *Arca noae* compound eye (Courtesy of U. Sauder). A) Transversal section through a compound eye showing a ciliary photoreceptor cell with surrounding pigment cells. B) Transversal section through the photoreceptive element showing the 9 x 2 + 2 arrangement of the cilia (Arrows). Abbreviations: N, nucleus; DMS, distal mitochondrial segment; PS, photoreceptive segment; PMS, posterior mitochondrial segment.

Electron micrographs show that each ommatidium is build up by a funnel-shaped tube of pigment cells with a photoreceptor cell at the bottom of the depression (Figure 1.2A). Above the photoreceptor cell, the empty space of the pigment funnel is filled by long microvilli. It can not be deduced from our electron micrographs whether these microvilli are parts of the photoreceptor cell or produced by the shielding pigment cells. However, a previous study investigating the compound eyes of *Barbatia*, suggests that the microvilli are build from the unpigmented, distalmost part of pigment cells surrounding the receptor cell (Nilsson, 1994).

The eye surface of *Arca* is covered by small extracellular vesicles.

The photoreceptor cell can be divided into four distinct regions along the proximodistal axis as already proposed for the compound eye of *Barbatia* (Nilsson, 1994) (see also Figure 1.2A). Most distally, the cell nucleus is located, followed by a region with numerous small mitochondria. Directly subjacent lies the photoreceptive segment and most proximal a few large mitochondria are found. At the proximal tip the axon emerges and joins the pallial nerve.

The photoreceptive element is build up by numerous cilia with enlarged membranes (Figure 1.2B). Each cilium provides several flattened sacks giving rise to the whole photoreceptive element composed of numerous piled membranes. Contrary to the cilia found in rods and cones of vertebrates which have a 9 x 2 + 0 microtubuli arrangement, *Arca* ciliary photoreceptors have a complete (9 x 2 + 2) set of microtubules (Figure 1.2B).

2. Arca noae Pax6 (AnPax6)

The finding that all three major eye-types are represented in the same phylogenetic class, the Bivalvia, raises the question about their phylogenetic relationship. Originally it was supposed that these eye-types evolved polyphyletically as new formations within the bivalvian class (Salvini-Plawen and Mayr, 1977). However, the finding of *Pax6* as a key regulator in eye morphogenesis throughout the animal kingdom suggests a monophyletic origin of the eye. Therefore, the investigation of *Pax6* is an excellent starting point to investigate the phylogenetic relationship between different eye-types.

2.1 Isolation of the *AnPax6* full length cDNA

In order to isolate a *Pax6* homolog from *Arca noae* a low-stringency PCR approach was carried out, using two degenerated primers directed against two highly conserved regions within the paired domain (YYETG and WEIRD). A 135 basepair long fragment, designated *AnPax6*-PCR1, was isolated from genomic DNA showing extensive sequence homology to other known *Pax6* homologs (Figure 2.1). At the amino acid level the fragment showed sequence homology of 90% to the corresponding part of mouse and human Pax6, 89% to eyeless and 88% to squid Pax6.

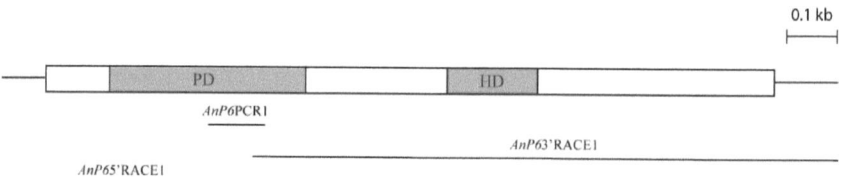

Figure 2.1 Schematic of the *AnPax6* full length cDNA. The open reading frame is boxed and the Paired domain and Homeodomain are indicated by a light grey and a dark grey box, respectively. The PCR clone, the 3'Race clone and 5'Race clone are shown underneath.

To get the complete sequence information of *AnPax6*, two sets of specific nested primers were generated for each direction, upstream and downstream, to carry out **RACE-PCR** (**R**apid **A**mplification of **c**DNA **E**nds). The primer sets correspond to the amino acid sequences RPRAIGGS and SKPRVATND of the paired domain. Messenger RNA was isolated from the eye-bearing mantle tissue using magnetic beads covalently linked oligo-dT tails (Dynabeads Oligo(dT)$_{25}$) Reverse transcription and PCR amplification was carried out by using a 3'RACE procedure. After a second round of nested PCR amplification, a fragment was isolated, designated *AnPax6-3'RACE*, which spans the 3'prime region of the paired box, the linker region between the paired box and the homeobox, the homeobox and the complete 3'-terminal region of *AnPax6* (Figure 2.1). To isolate the 5'upstream region of *AnPax6* a 5' RACE-PCR was carried out. A fragment was isolated, designated *AnPax6-5'RACE*, containing the 5' region of the paired domain, the coding sequence upstream of the encoded Paired domain and the 5'-untranslated region (Figure 2.1).

Using specific primers directed against the outermost 5'-end and 3'-end of *AnPax6* sequence, an uninterrupted full length cDNA of *AnPax6* was finally isolated.

2.2 Nucleotide and amino acid sequence of *AnPax6*

The *AnPax6* cDNA has a length of 1625 bp and a predicted open reading frame of 1416 bp encoding a protein of 472 amino acid residues (Figure 2.2.1). A putative initiator

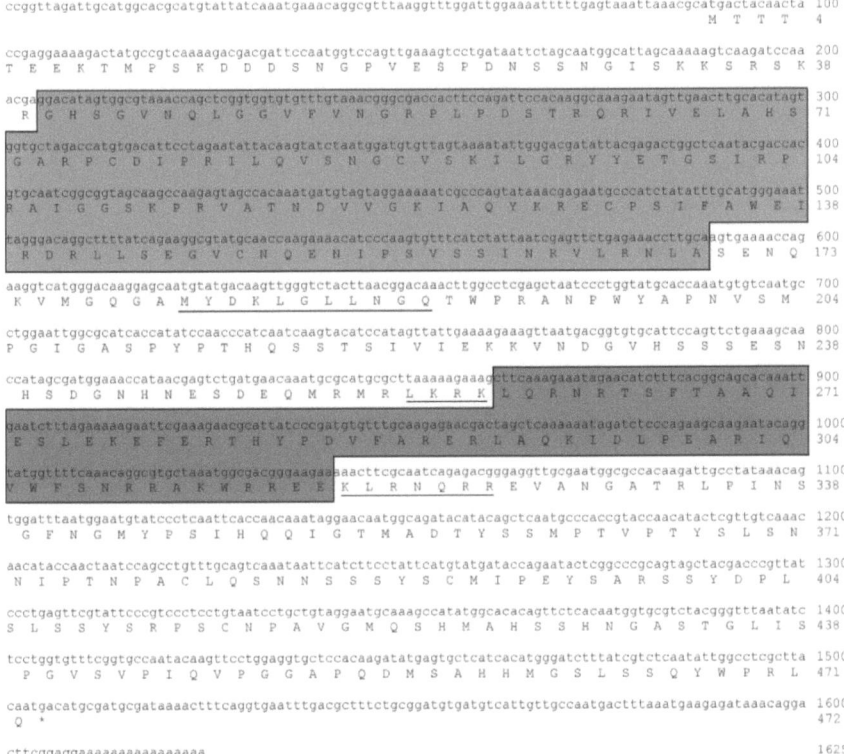

Figure 2.2.1 Nucleotide and deduced amino acid sequence of *AnPax6* cDNA. The paired domain is boxed in red, the homeodomain in green. The conserved linker region and the conserved amino acids flanking the homeodomain are underlined. The stop codon is indicated by an asterisk.

methionine codon was found 39 codons upstream of the encoded Paired domain. A termination codon was found at position 1504, 111 codons downstream of the encoded homeodomain. The paired domain has a length of 128 amino acid residues and is separated by a 91 amino acid long linker region from the 60 amino acid long homeodomain. The carboxy-terminal region has a length of 111 amino acids.

2.3 Sequence comparison of the Paired domain

At the amino acid level, the AnPax6 paired domain has 94% sequence homology to the corresponding domain of the mouse and human Pax6 (Ton et al., 1991; Walther and Gruss, 1991), 95% to zebrafish, nemertine (*Lineus*) and squid (*Loligo*) paired domain (Krauss

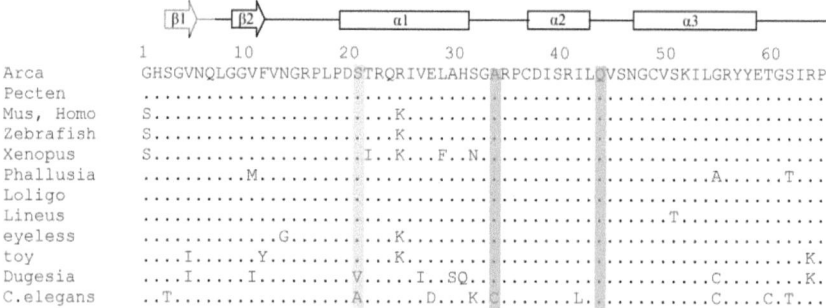

Figure: Comparison of the amino acid sequences between paired domains of different species. The secondary structure of the domain is shown at the top of the figure. The *Arca* sequence is shown in full; for other sequences only differing amino acids are shown. Dots represent identical amino acids. Shaded bars indicate Pax6-specific amino acid residues. Numbers behind the sequences give percent identity compared to *Arca* (first row).

et al., 1991; Loosli et al., 1996; Tomarev et al., 1997), 91% to EY and TOY paired domain (Czerny and Busslinger, 1995; Quiring et al., 1994) (Figure 2.3.1). The paired domain of AnPax6 is most similar to the paired domain of *Pecten maximus* Pax6 (97%). The flatworm *Dugesia* (Callaerts et al., 1999), which has a strongly diverged paired domain, and the nematode *C. elegans* (*vab-3*), (Zhang and Emmons, 1995) show significantly lower sequence homology to the paired domain of AnPax6 (80% and 84%, respectively).

The amino acid residues at position 42 (Isoleucine), 44 (Glutamate) and 47 (Asparagine) of the paired domain have been shown to determine Pax6 DNA binding specificity, with the asparagine at position 47 also found in other Pax proteins (Czerny and Busslinger, 1995). In *AnPax6*, all three corresponding amino acids are conserved (Figure 2.3.1). Moreover, all other Pax6 specific amino acid are conserved in AnPax6. Consistently, AnPax6 has a serine at position 21, alanine at position 34, arginine at position 66, cysteine at position 91, isoleucine at position 114 and an alanine at position 128.

Compared to human and mouse Pax6 paired domain, there are eight variations found in the amino acid sequence of AnPax6 (positions 1, 25, 78, 79, 82, 110, 111 and 112). At four positions AnPax6 differs from the majority of Pax6 paired domains. However, the amino acids at these positions are shared with at least one other Pax6 homolog. The asparagine at position 78 is shared with *Dugesia*, the aspartate at position 79 with *C. elegans*, the glycine at position 82 with *Lineus* and the glutamate at position 112 with *Xenopus*.

2.4 The linker region

The paired domain and the homeodomain of AnPax6 are linked by a 91 amino acid long linker region. Linker regions of known Pax proteins show only little sequence homology with the exception of a small eleven amino acid long motif, MYDKLGLLNGQ (Figure 2.2.1). This motif is highly conserved in vertebrate and most invertebrate Pax6 and is also found in AnPax6.

The conserved octapeptide which is found in all other groups except the Pax4/6 group, is lacking in the AnPax6-linker region. At the C-terminal end of the linker, immediately adjacent to the homeodomain, a stretch of four amino acids are highly conserved in Pax6 proteins (Figure 2.2.1).

2.5 Sequence comparison of the homeodomain

The homeodomain of AnPax6 shows sequence identity to other Pax6 homologs ranging from 75% to 98% (Figure 2.5.1). Amino acid identity of the homeodomain is 98% to the cephalopod mollusc *Loligo*, 95% to *Lineus*, 93% to the aligned vertebrates and *C. elegans*,

```
                            α1                α2            α3/4
              1       10        20        30        40        50        60    %id
Arca          LQRNRTSFTAAQIESLEKEFERTHYPDVFARERLAQKIDLPEARIQVWFSNRRAKWRREE     100
Pecten        ............................................................   100
Mus, Homo     ........QE...A.....................A........................    93
Zebrafish     ........QE...A.....................A........................    93
Xenopus       ........QE...A.....................A........................    93
Phallusia     ........SQE.V.A....................S........................    90
Loligo        .............A...............................................   98
Lineus        ........N....A...............................................   97
Eyeless       ........ND..D..........................G..G.................    92
toy           ........SNE..D.........................D..G.................    90
Dugesia       S..S....ND..NL..............S..K.S.NLKVA.T...................   75
C.elegans     .........V..S.............................Q.................    95
```

Figure 2.5.1 Comparison of the amino acid sequences between homeodomains of different species. The secondary structure of the domain is shown at the top of the figure. Shaded bars indicate the Pax6-specific amino acids. The *Arca* sequence is shown in full; for other sequences only differing amino acids are shown. Dots represent identical amino acids. Numbers behind the sequences give percent identity compared to *Arca* (first row).

92% and 88% to Eyeless and Toy homeodomains, respectively. Like for the paired domain, lowest sequence identity is found to the homeodomain of the planarian *Dugesia* (75%). Compared to human and murine Pax6 homeodomains, AnPax6 diverges at four positions (10, 11, 15 and 36,) (Figure 2.5.1).

In the third helix of the homeodomain, which is the DNA recognition domain, the AnPax6 amino acid sequence is homologues to all other known Pax6 homeodomains. Moreover, all amino acids known to be specific for Pax6 homeodomains are conserved in AnPax6 (Figure 2.5.1).

The carboxy-terminal region of AnPax6 has a length of 111 amino acid (Figure 2.2.1). Immediately adjacent to the carboxy-terminus of the homeodomain, AnPax6 has a seven amino acid long sequence motif (KLRNQRR, Figure 2.2.1), which is highly conserved and found in most other Pax6 homologs (Loosli et al., 1996).

The high sequence homology and particularly the conservation of Pax6 specific amino acid residues suggests that AnPax6 is indeed a *bona fide* Pax6 homolog.

2.6 Real-time PCR expression analysis of *AnPax6*

Real-time PCR is an extremely sensitive method to quantify low abundance messenger RNA. A big advantage of this method is that it is not necessary to quantify the concentrations of mRNA or cDNA in a sample before exposing it to real-time PCR. This technique was used because the first attempt to identify *AnPax6* expression in *Arca* compound eyes by *in situ* hybridization failed. Furthermore, real-time PCR is a helpful tool to screen multiple tissues for *Pax6* expression.

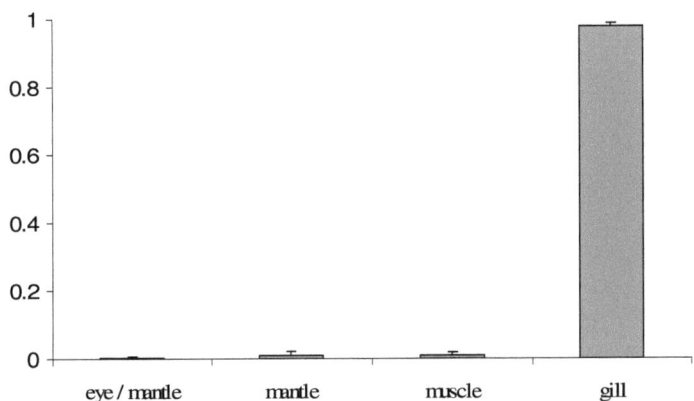

Figure 2.6.1 *Anpax6* gene expression analysis in different tissues. Graphs display relative values normalized to elongation factor expression levels.

To do so, mRNA was isolated from eye-bearing mantle tissue. Further, mRNA was isolated from mantle tissue without eyes as a control, from muscle tissue and from gill tissue. Isolated mRNA was reverse transcribed to single stranded cDNA by using random primers. To conduct real-time PCR, specific primers were designed for an amplicon size of 128 bp within the linker region of *Anpax6*. Real-time PCR expression analysis was done at least three times on independent cDNA templates, each generated from a different individual, using the Light cycler (Roche). As a reference to compensate for variations in quality and quantity of the preparations the housekeeping gene *Anef1α* (*Arca noae elongation factor 1α*) was used.

Real-time PCR data from *Arca noae* show only weak *AnPax6* expression in the eye-bearing mantle tissue (Figure 2.6.1), confirming the negative results obtained by *in situ* hybridization expression analysis. Also in the mantle and muscle tissues, *Anpax6* expression levels was found to be very low (Figure 2.6.1). Surprisingly, high *AnPax6* expression levels were found in the gills, about 300 times higher than in the eye-bearing mantle tissue. Because only relative expression values are obtained by real-time PCR, absolute expression levels are not deducible. However, the observation of high C_T values (threshold cycle, the first significant increase in the amount of PCR product) in the amplification plot (data not shown) suggests rather low expression levels. Moreover, the failure to confirm *AnPax6* expression in the gills by *in situ* hybridization also argues for low expression levels.

2.7 *AnPax6* is able to induce ectopic eyes in *Drosophila melanogaster*

Unfortunately, there are no tools available to study *Pax6* function in *Arca*. Transgenic methods have not been established for this organism and no mutants or appropriate cell lines are available. Thus, to have a rudimentary idea whether *AnPax6* might have the potential to function as an eye selector gene, it was tested whether it can induce ectopic eyes in *Drosophila*.

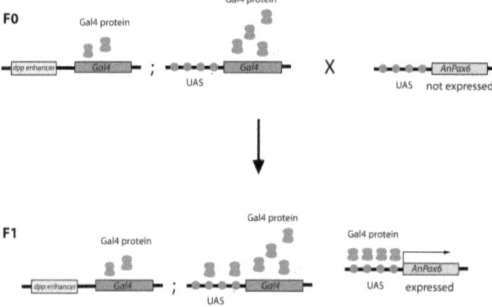

In the fly, ectopic eye formation on the wings, antennae and legs can be induced by ectopic expression of the *eyeless* cDNA in various imaginal discs (Halder et al., 1995) under the control of the GAL4-UAS system (Brand and Perrimon, 1993). GAL4 is yeast specific transcriptional activator which is able to drive

Figure 2.7.1 Targeted expression of *AnPax6*. F0: The driver line dpp^{brink} GAL4; UAS-GAL4 is crossed with the UAS-*AnPax6* line. F1: The progenies of this cross ectopically express *AnPax6* in various imaginal discs.

transcription of any gene of interest when introduced into the fly and if the gene of interest is preceded by several GAL4 upstream activating sequences (UAS).

Similar results were obtained by ectopic expression of the mouse, ascidian and squid *Pax6* that are all capable to induce ectopic eyes in the fly (Glardon et al., 1998; Halder et al., 1995; Tomarev et al., 1997). For this functional assay, *AnPax6* full length cDNA was inserted into the pUAST vector and injected into *yellow/white* strain embryos. Several UAS-*AnPax6* transgenic lines were generated. Since an initial attempt to generate ectopic eyes failed when using the dpp^{brink} Gal4 line, the dpp^{brink} Gal4; UAS-Gal4 line was used for stronger activation. Eight independent UAS-*AnPax6* lines were crossed to flies of the dpp^{brink} Gal4; UAS-Gal4 driver line (Figure 2.7.1).

In all generated crosses, ectopic eye structures were induced on legs and wings of adult flies (Figure 2.7.2). Scanning micrograph show distinct hexagonal facets and interommatidial bristles (Figure 2.7.2A and B). The ommatida of the ectopic compound eye show similar organization to that in the compound eye, although the interommatidial bristles are irregularly spaced.

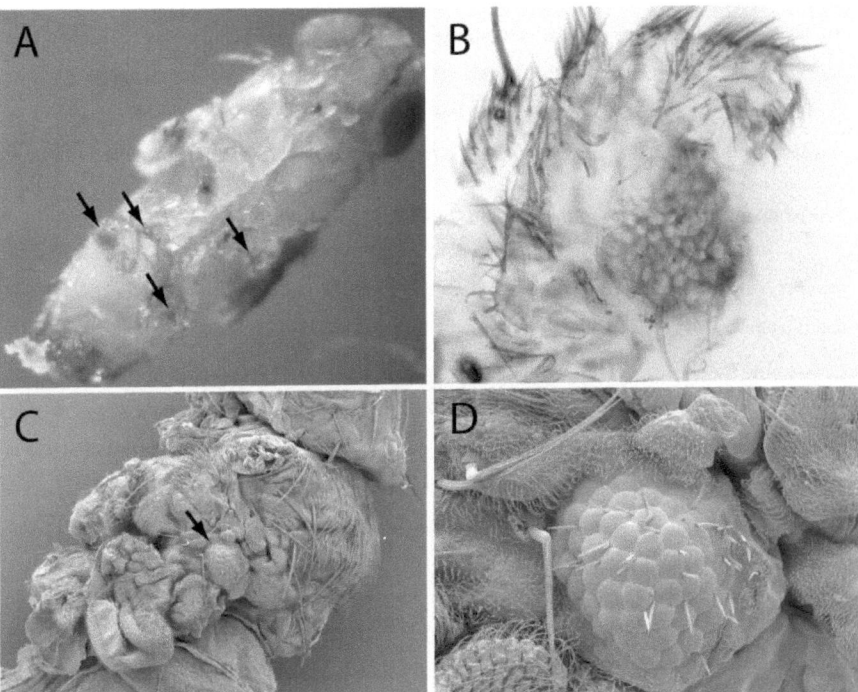

Figure 2.7.2 Ectopic eye structures in *Drosophila melanogaster* induced by overexpression of the bivalvian *AnPax6*. (**A, B**) Micrograph of ectopic eyes. (**C, D**) Scanning electron micrograph of ectopic eyes. (**A**) Overview of an adult fly with several ectopic eyes on the legs. (**B**) Dissected leg with a large outgrowth of eye tissue. (**C**) Overview of the fly. (**D**) Higher magnification of C. The ectopic eye has an array of hexagonal ommatidia and interommatidial bristles. Arrows point to ectopic eyes.

3. *Arca noae Six1/2 (AnSix1/2)*

The *Sine oculis* gene of *Drosophila*, which is a homolog of the Six1/2/so subfamily, was found to be involved in the eye regulatory network and to be a direct target of *ey* and *toy* (Niimi et al., 1999; Punzo et al., 2002). Consistently, the *so* homolog of planarians has been shown to be essential for eye regeneration (Pineda et al., 2000). In addition, the *Drosophila* Six3/6 homolog *optix* can induce ectopic eyes in an *eyeless*-independent way (Seimiya and Gehring, 2000). Coincidently, the vertebrate *Six3* and *Six6* are also expressed in the eye (Kobayashi et al., 1998; Loosli et al., 1999; Zuber et al., 1999) whereby at least *Six3* is suggested to be under direct control of *Pax6* (Ashery-Padan et al., 2000; Chow et al., 1999).

3.1 Isolation of the *Arca noae Six1/2 (AnSix1/2)* full-length cDNA

In search for *Arca noae* six genes, a PCR approach was conducted using degenerated primers corresponding to the highly conserved C-terminal end of the six domain (PRTIWD) and the conserved C-terminal region of the six homeodomain (TQVSNWF).

A fragment of 165 bp, designated *An*SixPCR1 (Figure 3.1.1), was isolated from eye/mantle-specific cDNA templates which showed high sequence homology to the corresponding part of other six proteins. Subsequently, the obtained fragment was extended by RACE using specific nested primers. Two overlapping fragments were isolated, designated *An*Six5'RACE1 and *An*Six3'RACE1, spanning the encoded protein sequence as well as the non-coding regions (Figure 3.1.1).

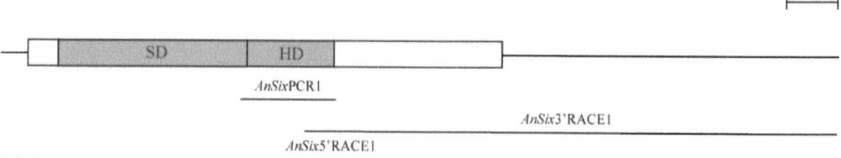

Figure 3.1.1 Schematic of the *AnSix1/2* full length cDNA. The open reading frame is boxed and the six domain and homeodomain are indicated by grey boxes. The PCR clone, the 3' RACE clone and 5' RACE clone are shown underneath.

3.2 Nucleotide and deduced amino acid sequence of *AnSix1/2*

The full length *AnSix1/2* cDNA has a length of 1629 bp encoding a 305 amino acid long protein (Figure 3.2.1). A putative initiator methionine was found 23 codons upstream of the encoded six domain. The 60 amino acid long homeodomain is immediately adjacent to the

115 amino acid long six domain. A termination codon is found 108 codons downstream of the six homeodomain at position 1072.

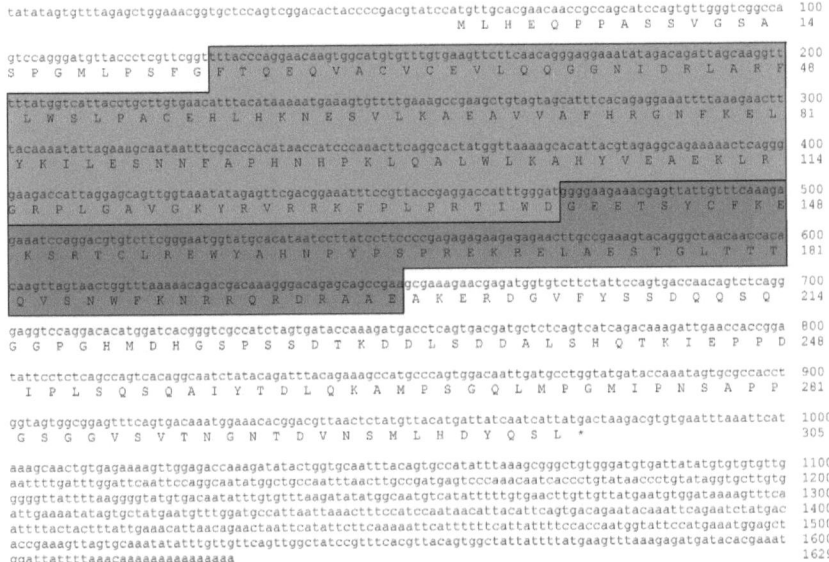

Figure 3.2.1 Nucleotide and deduced amino acid sequence of *AnSix1/2* full-length cDNA. The Six domain is boxed in red, the Six homeodomain is boxed in green.

3.3 The six domain

Within the six domain, AnSix1/2 shows highest sequence identity to the corresponding region of *Pecten maximus* Six1/2 (97% identity) and the polychaete (*Platynereis*) Six1/2 homolog (94% identity) (Figure 3.3.1). High sequence homology (around 90%) is also found to vertebrate Six1 and Six2. Significantly lower sequence identity is found to cnidarian (71%) and planarian (74%) Six1/2 homologs.

Two amino acids within the six domain of AnSix1/2 deviate from all other Six1/2 homologs (position 45 and 68). At the C-terminal end of the six domain, AnSix1/2 has a tetrapeptide (TIWD) which is also found in the vertebrate Six1 homolog and in Six1/2 homologs of invertebrates (Figure 3.3.1).

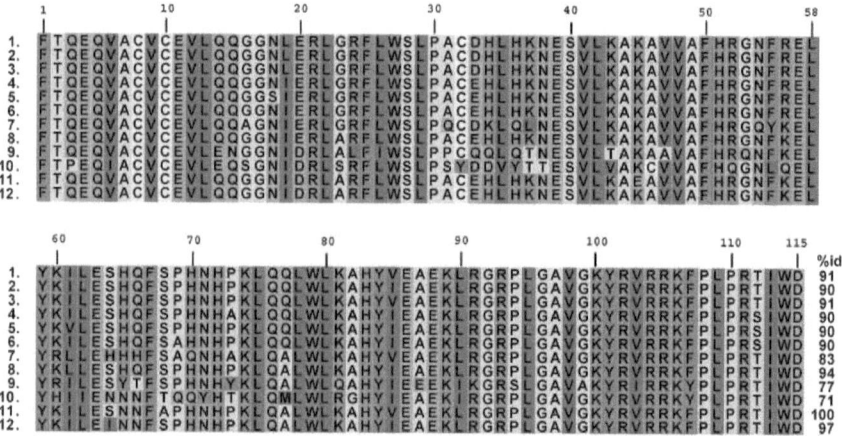

Figure 3.3.1. Comparison of the amino acid sequences between six domains of different species. The percentage of sequence identity to the respective AnSix1/2 amino acid sequence are indicated at the end of each line.

3.4 The six homeodomain

A diagnostic amino acid sequence specific for Six1/2 homeodomains is the tetrapeptide ETSY, which corresponds to positions 3 to 6 of the first helix within the homeodomain (Seo et al., 1999) and which is also found in AnSix1/2 (Figure 3.4.1). The homeodomain of AnSix1/2 shows highest sequence identity to the corresponding *Platynereis* Six1/2 homeodomain (97%; Figure 3.4.1), but is also highly homologous to the vertebrate six homeodomains of Six1 and Six2 (92-95%). *Cladonema* and *Dugesia* Six1/2 homologs display the most diverged homeodomain sequence (85% and 88% identity, respectively) compared to that of AnSix1/2 (Figure 3.4.1). The sequence homology among the two bivalvian Six1/2 homologs is rather low within the homeodomain (92%) compared to the high sequence identity found to other Six1/2 homologs of less related species (Figure 3.4.1). AnSix1/2 and

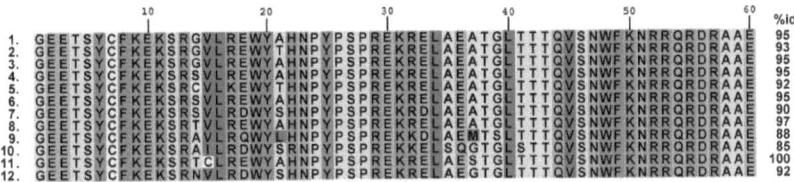

Figure 3.4.1 Sequence comparison of the six homeodomains of different species. The percentage of sequence identity to the respective AnSix1/2 amino acid sequence are indicated at the end of each line.

PmaSix1/2 differ at five positions (14, 15, 18, 21 and 37) within the homeodomain (Figure 3.4.1). However, the amino acid sequence is highly conserved in the recognition helix 3 at the C-terminal end of the homeodomain.

The C-terminal region comprises 107 amino acids rich in serine (15%) and proline (9.3%), suggesting the presence of transactivation functions (Figure 3.2.1). Numerous amino acid doublets are found in the C-terminal region of the protein, a feature also found in other six proteins (Pineda et al., 2000) for which the significance is not known.

The high sequence conservation within the six domain and the homeodomain, together with the phylogenetic analysis of AnSix1/2 strongly suggest it to be a member of the Six1/2 family.

3.5 Real-time PCR expression analysis of *Ansix1/2*

In *Drosophila*, *so* was shown to be expressed in the optic lobe primordia anterior to the cephalic furrow (Cheyette et al., 1994) Later, *so* expression is detected in the eye discs on both sides of the morphogenetic furrow. Just in front of the furrow, *so* is expressed within the undifferentiated cells of the eye disc primordium, whereas posterior to the furrow expression becomes restricted to individual photoreceptor cell clusters (Cheyette et al., 1994). Furthermore, *so* is also expressed in the ocelli region of the eye disc.

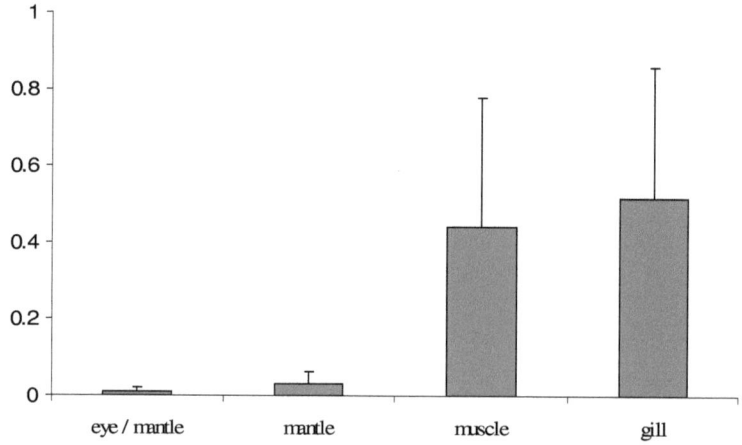

Figure 3.5.1 *Ansix1/2* expression analysis in different tissues of *Arca*. Graphs display the relative values normalized to elongation factor expression levels.

In vertebrates, *Six1* was found to be expressed in the otic and olfactory placode (Laclef et al., 2003b; Zheng et al., 2003) and is also suggested to play an important role in myogenesis (Laclef et al., 2003a). Coincidently, *Six1/2* of the cnidaria *Cladonema* is expressed in the subumbrellar striated muscle and *Six1/2* expression was also observed in nerve cells of *Cladonema* (Stierwald et al., 2004).

To investigate *Ansix1/2* expression in various tissues of *Arca*, a real-time PCR analysis was carried out. Messenger RNA was isolated from eye-bearing mantle tissue, mantle tissue without eyes, muscle tissue and gill tissue and reverse transcribed using random primers to generate cDNA. Specific primers to conduct real-time PCR were designed which generate a 118 bp long amplicon. Real-time PCR expression analysis was carried out three times independently using different cDNA templates isolated from three different individuals. As a reference for normalization and to compensate variations in the quality and the quantity of different cDNA preparations the housekeeping gene *Anef1α* was used.

In the eye-bearing mantle tissue and the mantle tissue, only low level of *AnSix1/2* expression was observed (Figure 3.5.1). In contrast, much higher expression levels were found in the muscle and the gill tissues (Figure 3.5.1).

No *AnSix1/2* expression was detected in the muscle and gill tissues by *in situ* hybridization. However, the high C_T-values indicated by the real-time PCR amplification plot (data not shown) suggest low expression levels, most likely below the threshold levels needed to detect *in situ* hybridization signals.

4. Arca naoe opsin gene (AnOpsinX)

A peculiarity of the *Arca* compound eye is the use of ciliary photoreceptor cells, in contrast to most other compound eyes that employ rhabdomeric photoreceptor cells. Theses two photoreceptor cell types not only differ in their morphological fine structure but also in the employment of different types of opsins. Phylogenetic analysis clearly groups the opsins in two clusters: ciliary photoreceptor cells, dominantly found in vertebrate eyes employ c-opsins, whereas rhabdomeric photoreceptor cells which are generally found in compound eyes employ r-opsins (Arendt et al., 2004). The fact that the *Arca* compound eye represents the atypical case employing ciliary photoreceptor cells raises the question about which opsin type is used in these compound eyes.

4.1 The isolation of *AnOpsinX* full-length cDNA

In search for opsin genes in *Arca* a low-stringency PCR approach was carried out, using degenerated primers raised against two highly conserved regions within the fifth and seventh transmembrane helix of opsins. Eye-bearing mantle tissue was used to isolate mRNA and to generate eye/mantle tissue-specific cDNA templates. A 474bp long fragment, designated AnOpsPCR1, was isolated (Figure 4.1.1). The fragment showed 33% amino acid sequence identity to the Go-coupled rhodopsin of the sea urchin *Strongylocentrotus* and 29% to the Go-coupled rhodopsin of the pacific scallop *Patinopecten*.

To obtain the full cDNA sequence, the fragment was extended by RACE using specific primers. Two overlapping fragments spanning the coding and untranslated regions were isolated, designated *An*Ops5'RACE1 and *An*Ops3'RACE1 (Figure 4.1.1).

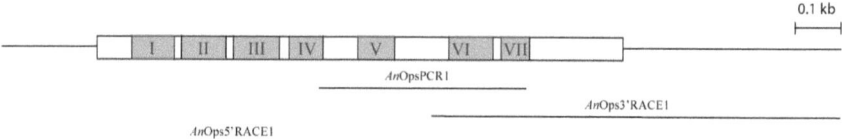

Figure 4.1.1 Schematic of the *AnOpsinX* full length cDNA. The open reading frame is boxed and the transmembrane domains are indicated by grey boxes which are serially numbered. The PCR clone, 3' RACE and 5' RACE clones are shown underneath.

4.2 Nucleotide and deduced amino acid sequence of *AnOpsinX*

The full length *AnOpsinX* cDNA has a length of 1865 bp with a putative open reading frame of 1179 bp coding for a 390 amino acid long protein (Figure 4.2.1). A putative translation initiation site is found 30 codons upstream of the encoded helix I and the termination codon is located at position 1382, 73 codons downstream of helix VII. The C-terminal region has a length of 72 amino acids.

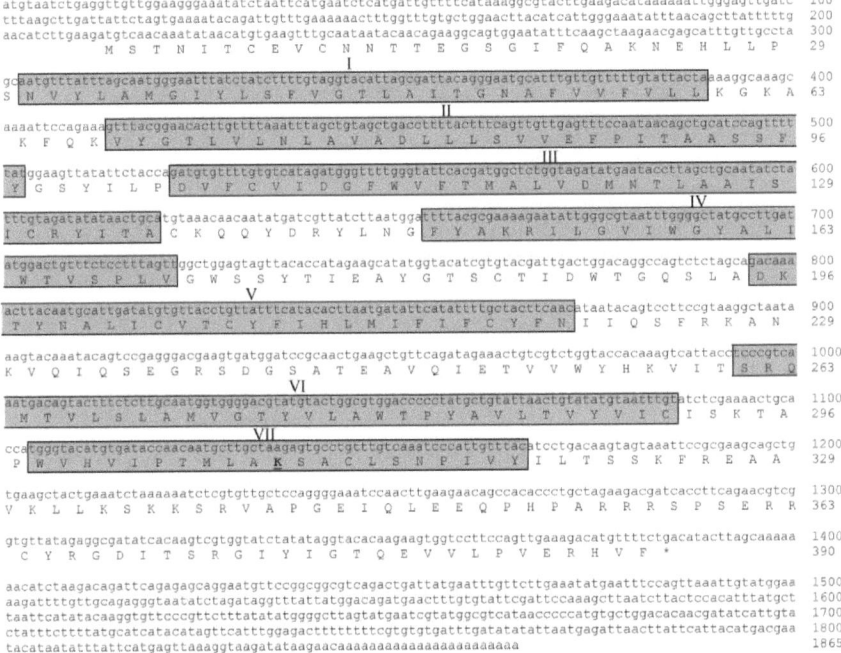

Figure 4.2.1 Nucleotide and deduced amino acid sequence of the *AnOpsinX* full-length cDNA. Boxed amino acid sequences (red) correspond to the transmembrane domains serially numbered from I to VII. The highly conserved lysine of helix VII is underlined.

Seven transmembrane domains were identified by multiple alignments of opsins and by comparison to the well studied two-dimensional structure of bovine rhodopsin (Palczewski et al., 2000). Importantly, the highly conserved lysine residue which covalently links the chromophore through a Schiff base to the opsin protein is also found in the seventh transmembrane domain of AnOpsinX (Figure 4.2.1).

4.3 Structural analysis of AnOpsinX

The hydropathy plot of AnOpsinX suggests several hydrophobic segments (Figure 4.3.1). In particular, six hydrophobic segments are obvious from the blot, consistent with six of the seven putative transmembrane helices. The seventh helix does not show well in the hydropathy analysis, most probably because it is quite polar in the middle with one lysine (K308) and two serines (S309 and S313).

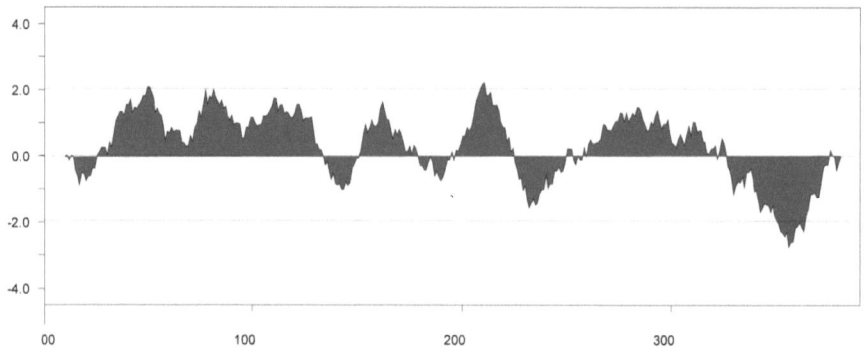

Figure 4.3.1 Hydropathy analysis of AnOpsinX. Six hydrophobic segments are apparent in the plot, corresponding to six of total seven transmembrane domains.

A two dimensional model of AnOpsinX (Figure 4.3.2) was generated based on multiple sequence alignment and comparison to the two-dimensional model of bovine rhodopsin (Palczewski et al., 2000). This model offers a structural template for other G-protein coupled receptors (GPCR) since the length of the seven transmembrane helices and the three extracellular loops are expected to be nearly the same for most of the family members (Palczewski et al., 2000). Variations are mostly found in the intracellular loops, particularly in the third cytoplasmic loop which is critical in the binding of G-protein, arrestin and rhodopsin kinase, and probably reflects the specificity of the receptor.

The asparagine residue at position 50 (N55 in bovine rhodopsin; see also Figure 7.1.1 of section 7.1 in the Introduction) of helix I is highly conserved among related proteins and was shown to be responsible for the interhelical hydrogen bond to D79 (D83 in bovine rhodopsin) another conserved amino acid in helix II (Figure 4.3.2).

The first cytoplasmic loop of GPCRs generally has several positively charged amino acid residues which may be critical for proper insertion into the membranes (Hartmann et al., 1989), a feature also found in AnOpsinX (K59, K63 and K66).

At the extracellular end of helix III a highly conserved cysteine is found at position 106 (C110 in bovine rhodopsin), which is likely involved in a disulfide bond with a second cysteine found at position C183 (Figure 4.3.2).

In bovine rhodopsin, E113 in helix III is the counterion for the protonated Schiff base formed between the retinal and K296. In AnOpsinX, the corresponding position for the counterion is occupied by an aspartate (D109; Figure 4.3.2).

All opsins contain a highly conserved E(D)RY motif at the C-terminal end of helix III, which have been implicated in the regulation of the receptor's interaction with its G-protein. In AnOpsinX, a cysteine (C130) instead of a glutamate or an aspartate is found at the first position, whereas the arginine (R131) and the tyrosine (Y132) residues of the motif are

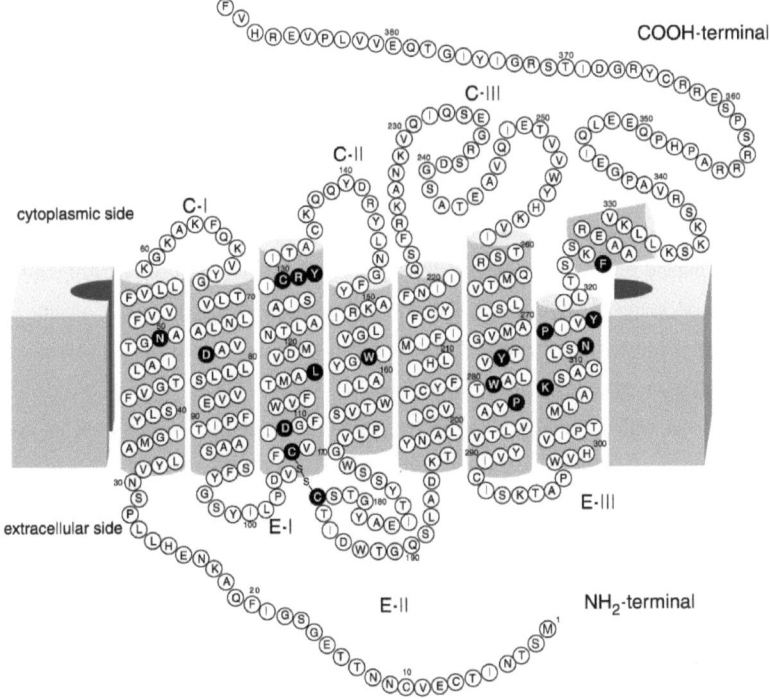

Figure 4.3.2 Two dimensional model of AnOpsinX. Modified after (Hartmann et al., 1989). Some of the key residues which are conserved in opsin proteins are shown in filled circles.

conserved (Figure 4.3.2).

The third cytoplasmic loop, C-III, is critical in the binding of G-proteins, arrestin and rhodopsin kinase and is highly variable among different GPCRs. Compared to bovine rhodopsin, C-III of AnOpsinX is considerably longer, probably reflecting a deviated function and specificity in G-protein binding.

Four highly conserved residues, F261, W265, Y268 and A269 were found to surround the β-ionone ring of retinol in bovine rhodopsin. Three of these four amino acids, W279, Y282 and A283 are conserved at the corresponding positions of AnOpsinX (Figure 4.3.2). Helix VII is critical for rhodopsin function because it contains the lysine (K296 in bovine rhodopsin) which forms the retinylidene linkage with the chromophore. In AnOpsinX the corresponding lysine is found at position 299 (Figure 4.3.2).

The C-terminal end of helix VII contains a highly conserved NPXXY motif which is found in all GPCRs, corresponding to N305, P306, I307, V308 and Y309 in AnOpsinX (Figure 4.3.2).

4.4 Real-time PCR expression analysis of *AnOpsinX*

Because no *AnOpsinX* expression could be detected in the compound eye of *Arca* by *in situ* hybridization, the possibility that *AnOpsinX* may have a non-visual function expressed also outside of the eye was considered.

To pursue this assumption, a real-time PCR expression analysis of *AnOpsinX* was performed. Messenger RNA was isolated from eye-bearing mantle tissue, mantle tissue without eyes, adductor muscle and gill tissues and reverse transcribed using random primers to generate tissue-specific cDNA templates. For real-time PCR, specific primers were designed that generate amplicons of 116 bp. Three independent real-time PCR experiment were carried out. For each experiment, a different set of cDNA templates isolated from a different individual was used. To compensate for differences in the quality and quantity of the cDNA samples and for normalization, the *Arca noae* specific housekeeping gene *Anef1α* was used.

Consistent with our *in situ* hybridization data, *AnOpsinX* was found to be expressed only at low expression levels in the eye-mantle tissue (Figure 4.4.1). Much higher expression levels were found in the gills and the adductor muscle (Figure 4.4.1). Although much higher *AnOpsinX* expression levels are found in muscle and gill tissues no expression was detected by *in situ* hybridization. However, the high C_T-values indicated by the real-time PCR amplification plot (data not shown) suggests that *AnOpsinX* expression might be below the threshold levels to detect transcripts by *in situ* hybridization.

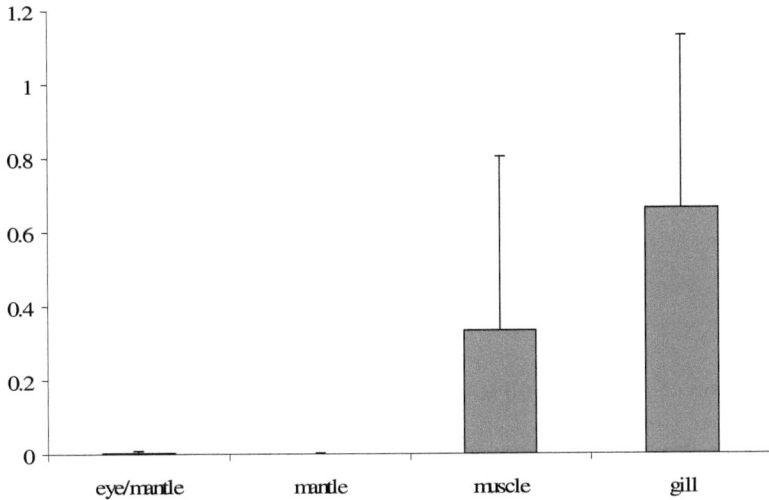

Figure 4.4.1 Expression analysis of *AnOpsinX* by real-time PCR. Graphs display the relative values normalized to elongation factor expression level.

IV. Results (Pecten)

1. Pecten Pax6 (PmaPax6)

Pecten maximus represents the mirror eye, a further eye-type found in bivalves which was investigated. As a starting point to study the evolutionary relationship between the compound eye of *Arca* and the mirror eye of *Pecten*, a *Pax6* homolog was isolated from both species.

1.1 Isolation of PmaPax6 full length cDNA

To isolate the *Pax6* homolog of *Pecten*, the same PCR approach was used as previously described for the isolation of *AnPax6*. Using degenerated primers directed against two regions highly conserved in all known paired domains corresponding to the amino acid sequences YYETG and WEIRD, a 135 bp long fragment was isolated, designated *PmaP6*PCR1 (Figure 1.1.1), from genomic DNA. The isolated fragment showed more than 90% sequence identity at the amino acid level to the corresponding part of the paired domain of other Pax6 proteins.

Subsequently, a set of specific nested primers were designed corresponding to the amino acid sequence RAIGGSKPRV and AQYKRECPS to extend the fragment by RACE-PCR. For that purpose, mRNA was extracted from excised eyes and reverse transcribed to cDNA templates. By 3'RACE-PCR a fragment was isolated, designated *Pma*P63'RACE1 with a length of 1233 bp (Figure 1.1.1). The 5'-end of *PmaPax6* was isolated by a 5'RACE-PCR yielding a fragment, designated *Pma*P63'RACE1, comprising the amino-terminal part of

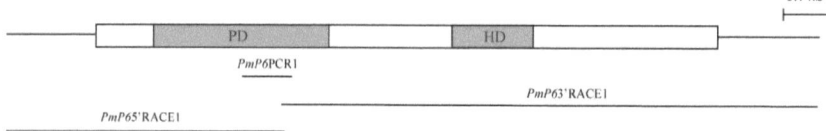

Figure 1.1.1: Schematic of *Pma*Pax6 full length cDNA. The open reading frame is boxed and the Paired box are indicated by grey boxes. The PCR clone and the RACE-PCR clones are shown underneath

the paired domain, the encoded protein sequence upstream of the paired domain and the 5' untranslated region. By using gene-specific primers directed against the outermost regions of PmaPax6 a full-length PmaPax6 clone was finally isolated.

1.2 Nucleotide and deduced amino acid sequence of *PmaPax6* full-length cDNA

The isolated *PmaPax6* cDNA has a length of 1915 bp and a predicted open reading frame of 1422 bp encoding a protein of 474 amino acid residues (Figure 1.2.1). A putative initiator methionine is located 42 codons upstream of the encoded paired domain. The termination codon is found at position 1631, 156 codons downstream of the encoded homeodomain. The paired domain has a length of 128 amino acids and is linked to the 60 amino acid long homeodomain by a 89 amino acid long linker region. The C-terminal region of PmaPax6 has a length of 155 amino acids.

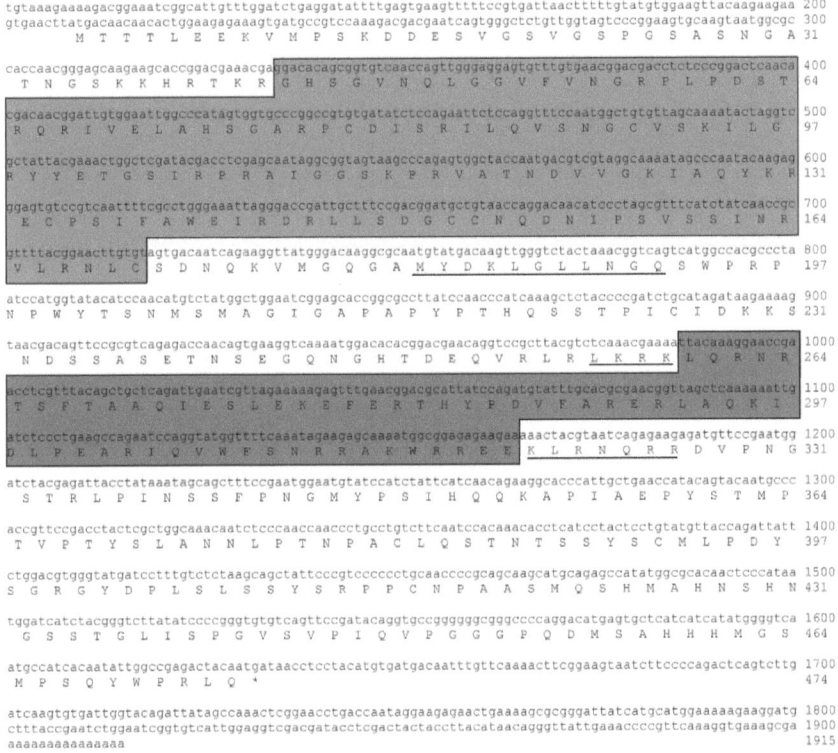

Figure 1.2.1 Nucleotide and deduced amino acid sequence of *PmaPax6* cDNA. The paired domain is boxed in red, the homeodomain in green. The PmaPax6 linker region contains a MYDKLGLLNGQ motif (underlined) that is highly conserved in most Pax6 linker regions.

1.3 Sequence comparison of the paired domain

The paired domain of PmaPax6 shows 91% sequence identity to the corresponding part of the mouse and human Pax6 (Ton et al., 1991; Walther and Gruss, 1991), 92% identity to the zebrafish Pax6 paired domain (Krauss et al., 1991) and 89% identity to *Phallusia* Pax6, Ey and Toy paired domain (Czerny et al., 1999; Glardon et al., 1997; Quiring et al., 1994)(Figure 1.3.1). High sequence homology is found to the Pax6-specific paired domains of the closely related lophotrochozoan species *Lineus* (94% sequence identity) (Loosli et al., 1996) and *Loligo* (93% sequence identity) (Tomarev et al., 1997)). Consistent with the close phylogenetic relationship of *Arca* and *Pecten*, the paired domains of AnPax6 and PmaPax6 show high sequence identity (97%). Compared to human and mouse the PmaPax6 paired domain deviates at eleven positions (1, 25, 78, 79, 82, 106, 108, 110, 111, 112 and 128). However, all Pax6-specific amino acids are conserved within the paired domain of PmaPax6,

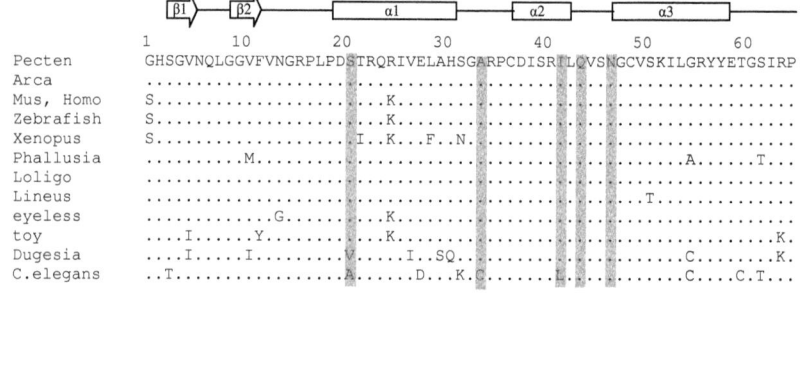

Figure 1.3.1 Comparison of the amino acid sequences between paired domains of different species. The secondary structure of the domain is shown at the top of the figure. The shaded bars indicate the Pax6 specific amino acids. The *Pecten* sequence is shown in full; for other sequences only differing amino acids are shown. Dots represent identical amino acids. Shaded bars indicate Pax6-specific amino acid residues. Numbers behind the sequences give percent identity compared to *Pecten* (first row).

with the sole exception of a cysteine which substitutes an alanine at the C-terminal end. Compared to AnPax6, PmaPax6 differs only at four positions (106, 108, 112 and 128). The two cysteines at position at position 108 and position 128 are not found in any other Pax6 homolog and therefore seem to be *Pecten*-specific.

1.4 The linker region

The linker region of PmaPax6 has a length of 89 amino acid residues. The highly conserved sequence motif (MYDKLGLLNGQ) found in most Pax6 homologs is also conserved in PmaPax6 (Figure 1.2.1). Moreover, a conserved amino acid stretch (LKRK) is found at the C-terminal end of the linker region (Figure 1.2.1).

1.5 Sequence comparison of the homeodomain

The homeodomain of *PmaPax6* and the *Pax6* genes from other species differ at positions in the first two α-helices and in the turn of the helix-turn-helix motif, whereas the recognition helix is identical (Figure 1.5.1). Moreover all Pax6-specific amino acids are conserved within the homeodomain of PmaPax6.

At the amino acid level, the homeodomain of PmaPax6 is 100% identical to the homeodomain of AnPax6. Very high amino acid sequence identity is also found to the Pax6 homeodomains of the lophotrochozoan species *Loligo* and *Lineus* (98% and 97%, respectively).

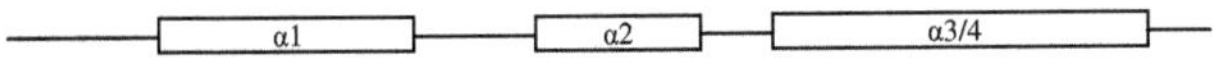

```
                       α1              α2              α3/4
                1     10        20        30        40        50        60    %id
Pecten          LQRNRTSFTAAQIESLEKEFERTHYPDVFARERLAQKIDLPEARIQVWFSNRRAKWRREE    100
Arca            ............................................................    100
Mus, Homo       ......QE...A................A...............................     93
Zebrafish       ......QE...A................A...............................     93
Xenopus         ......QE...A................A...............................     93
Phallusia       ....SQE.V.A.................S................................     90
Loligo          ...........A.................................................     98
Lineus          ......N....A..................................................    97
Eyeless         .....ND..D..................G..G..............................    92
toy             .....SNE..D.................D..G..............................    90
Dugesia         S..S...ND..NL............S..K.S.NLKVA.T.......................    75
C.elegans       .....V..S.........................Q...........................    95
```

Figure 1.5.1 Comparison of the amino acid sequences between homeodomains of different species. The secondary structure of the domain is shown at the top of the figure. Shaded bars indicate the Pax6-specific amino acids. The *Arca* sequence is shown in full; for other sequences only differing amino acids are shown. Dots represent identical amino acids. Numbers behind the sequences give percent identity compared to *Arca* (first row).

The carboxy-terminal region of PmaPax6 has a length of 136 amino acids, 25 amino acids more than the carboxy-terminal region of AnPax6 (Figure 1.2.1 and Figure 2.2.1 of section 2.1 in Results chapter III.). Just adjacent to the homeodomain PmaPax6 has a seven amino acid long motif (KLRNQRR), which is also found in most other Pax6 homologs (Figure 1.2.1).

1.6 Phylogenetic analysis of bivalvian AnPax6 and PmaPax6

A phylogenetic tree was generated by the neighbour joining method using the full-length paired domain, full-length homeodomain and flanking regions of Pax6 proteins as a basis for analysis (Figure 1.6.1).

Within the tree, *AnPax6* and *PmaPax6* cluster together, confirming the close phylogenetic relationship of *Arca noae* and *Pecten maximus*. Moreover, both *Pax6* genes cluster within the lophotrochozoan clade.

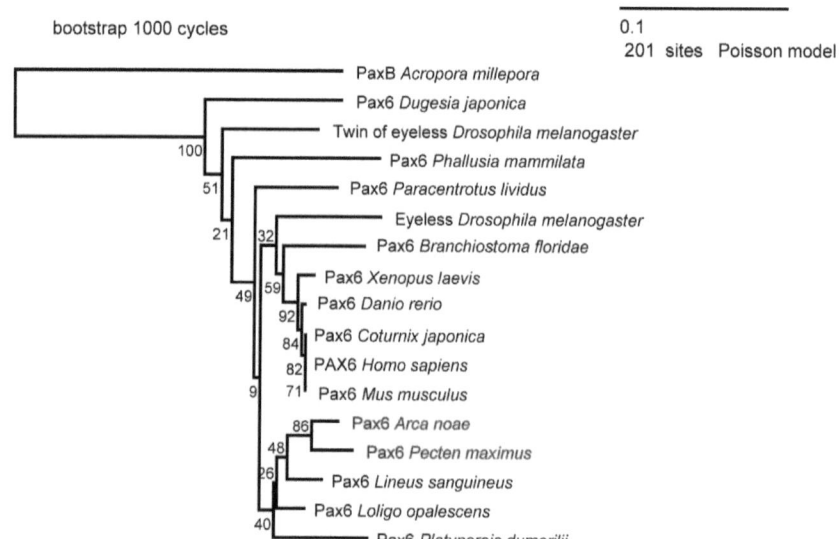

Figure 1.6.1 Phylogenetic analysis of *A. noae* and *P. maximus* Pax6 proteins. The phylogenetic tree confirms the close relationship between AnPax6 and PmaPax6 (both highlighted in red) and their close relationship to the lophotrochozoan clade. The tree was generated by the neighbour joining method using 201 sites from the paired domain, homeodomain and flanking regions. Bootstrap percentage values (1000 replicates) are indicated over the corresponding nodes.

1.7 Real-time PCR expression analysis of *PmaPax6*

Similar to *AnPax6*, it was not possible to detect *PmaPax6* expression in the *Pecten* mirror eye by *in situ* hybridization. To be sure that there is indeed no *PmaPax6* expression in the eye, a real-time PCR analysis was carried out, which is far more sensitive than *in situ* hybridization. To generate cDNA templates for real-time PCR, mRNA from eyes, muscle tissue, mantle tissue, ovary and gill tissue was isolated. Reverse transcription was performed by using random primers. Specific primers were designed that generate amplicons of 131 bp within the paired domain of *PmaPax6*. Three independent real-time PCR experiments were carried out. For each experiment a different set of cDNA templates was used, each isolated from a different individual. To compensate for differences in the quality and quantity of the cDNA samples and for normalization the *Pecten*- specific housekeeping gene, *Pmaef1α* was used.

Only weak *PmaPax6* expression was detectable in the eye by real-time PCR, explaining our negative results obtained by *in situ* hybridization (Figure 1.7.1). Also in the muscle tissue, mantle tissue and ovary tissue *PmaPax6* is only weakly expressed (Figure 1.7.1). Interestingly, *PmPax6* expression was found to be significantly higher in the gills than in any other tissue investigated (Figure 1.7.1). This is consistent with our previous observation that *Anpax6* is also much higher expressed in the gills.

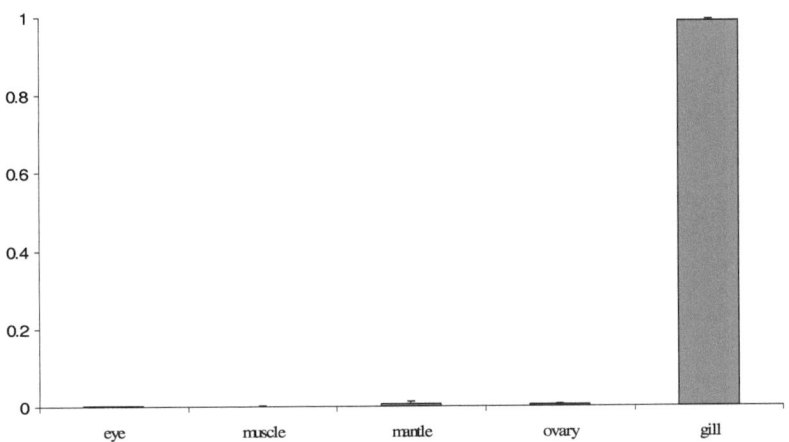

Figure 1.7.1 *PmaPax6* expression analysis in different tissues. Graphs display the relative values normalized to elongation factor expression levels.

1.8 Expression analysis of *PmaPax6* in *Pecten* larvae by whole mount *in situ* hybridization

No *PmaPax6* expression was detected in the eyes of adult *Pecten* by *in situ* hybridization. However, it was taken into consideration that *Pax6* may be important for eye morphogenesis during earlier developmental stages of the animal. Unfortunately, serious problems to find a source for early developmental stages (e.g. veliger larvae) of *Pecten* were encountered. A big disadvantage is, for example, that it is impossible to collect *Pecten* larvae in their natural environment. First of all, because spawning is seasonal and the time-point varies from place to place, making it utterly impossible to anticipate the accurate spawning time. Secondly, veliger larvae are very small (60-250µm) and are difficult to distinguish between other bivalvian species.

In principle, artificial spawning is possible but needs serious conditioning of the animals, impossible to accomplish without accurate facilities. Moreover, *Pecten* veliger larvae have to be nourished with a combination of various algae. Another serious problem are bacterial infections which frequently contaminate the entire breeding. Therefore, without the appropriate facility it was beyond our power to rear *Pecten* larvae.

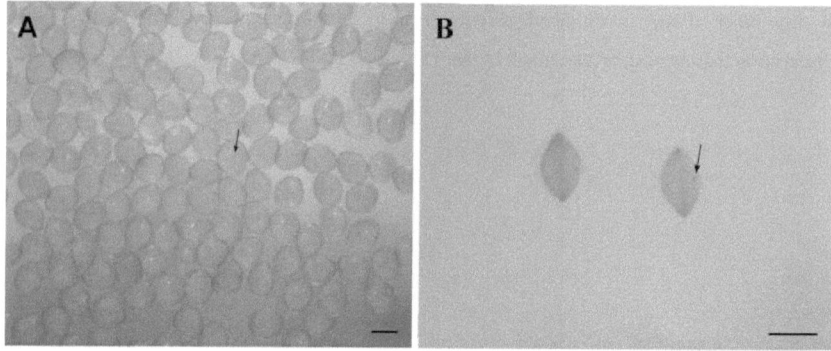

Figure 1.8.1 Bilateral larval eyes of *Pecten* veliger. (**A**) Lateral views of veliger larvae indicate one of the two bilateral eyes (**B**) whereas ventral views of veliger larvae indicate both eyes. Scale lines, 200µm.

After numerous attempts it was finally possible to get access to *Pecten* veliger larvae thanks to IFREMER in Brest (France) as they set up an experiment to study optimal rearing conditions. Unfortunately the investigations were restricted to a limited set of developmental stages because larval development takes long and there was only a short time-frame available to collect larvae. Therefore, the sampling was limited to the time-point where larval eyes start to appear (eyed veliger, Figure 1.8.) up to pediveliger stages just prior to metamorphogenesis (day 16 to day 27 after fertilization; for further discussions designated as stage 16 to 27). So far little is known about larval eyes of scallops, but it is assumed that they consist of one pigmented cell and a photoreceptor cell (Hodgson and Burke, 1988).

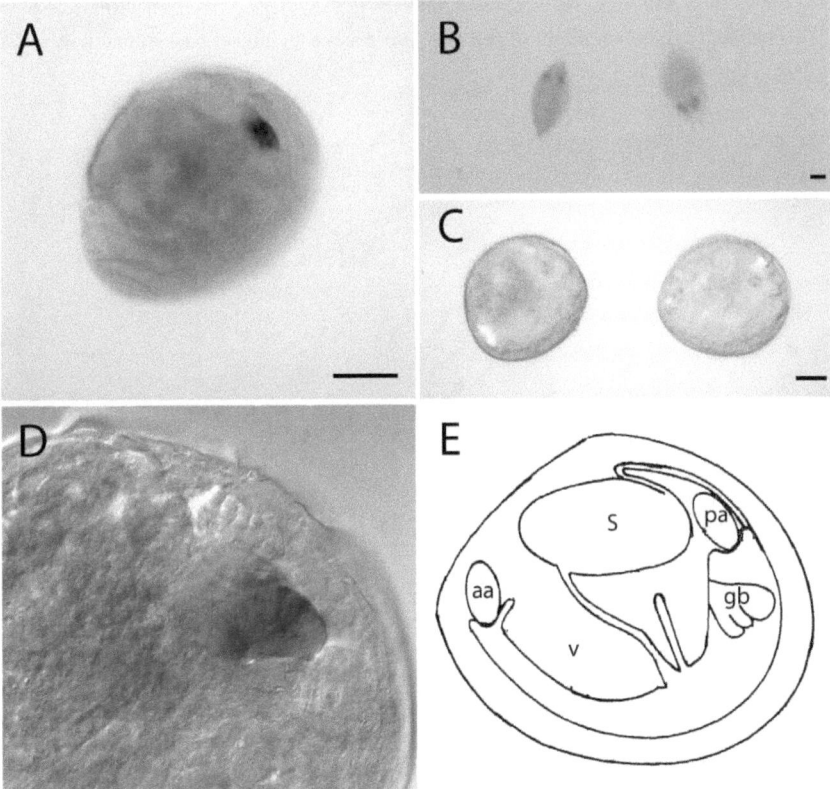

Figure 1.8.2 *In situ* hybridization on veliger larvae (stage 18) using DIG-labeled probes for *PmaPax6*. (**A**) Lateral view indicating *PmaPax6* expression in the gill primordia. (**B**) Ventral view demonstrating *PmaPax6* expression in bilateral gill primordia. (**C**) Negative control with Sense probe. (**D**) Close-up of A (Nomarski) indicating the gill bud. (**E**) Schematic of Pecten veliger larva (from Beninger et al., 1994). aa, anterior adductor muscle; gb, gill bud; pa, posterior adductor muscle; s, stomach; v, velum. Scale lines, 50µm.

At stage 27, the larval eyes are well visible under the binocular (Figure 1.8.1). At earlier stages however, weakly pigmented cells within the eye regions were only visible at higher magnification by light microscopy.

To investigate *Pax6* expression in *Pecten* veliger larvae, whole mount *in situ* hybridization was carried out using a DIG-labeled *PmaPax6* antisense probe on veliger larvae from stage 16 to stage 27.

No staining in the larval eyes was detected, but surprisingly staining in the gill primordia was observed(Figure 1.8.2).

PmaPax6 expression was found to be very week at stage 16 and 17 (data not shown) and started to be expressed more strongly in larvae stages 18 to 24. At subsequent stages (25-27), *PmaPax6* expression diminish and gets also masked by higher background levels (data not shown).

1.9 Targeted expression of PmaPax6 in Drosophila melanogaster

Similar to the situation in *Arca*, it is not possible to study *Pax6* function in *Pecten* due to the lack of mutants, transgenic techniques and appropriate cell lines. Therefore, it was tested whether *PmaPax6* has the potential to induce ectopic eyes in *Drosophila*.

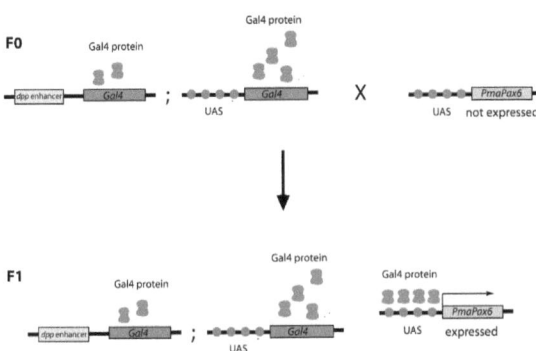

PmaPax6 full length cDNA was inserted into the pUAST vector and injected into *yellow/white* strain embryos. Several UAS-*PmaPax6* transgenic fly lines were generated and subsequently eight independent UAS-*PmaPax6* lines were crossed to flies of the driver line dpp^{brink} GAL4; UAS-GAL4 (1.9.1). All eight crosses resulted in progenies with ectopic eye structures on the wings and the legs (1.9.2). Scanning electron micrographs show the fine structure of the ectopic eyes (1.9.2C and D). The facets are well organized and also interommatidial bristles are seen (Figure 1.9.2D), although their spacing is quite irregular.

Figure 1.9.1 Targeted expression of *AnPax6*. **F0:** The driver line dpp^{brink} GAL4; UAS-GAL4 is crossed with the UAS-*AnPax6* driver line. **F1:** The progenies of this cross ectopically express *AnPax6* in various imaginal discs.

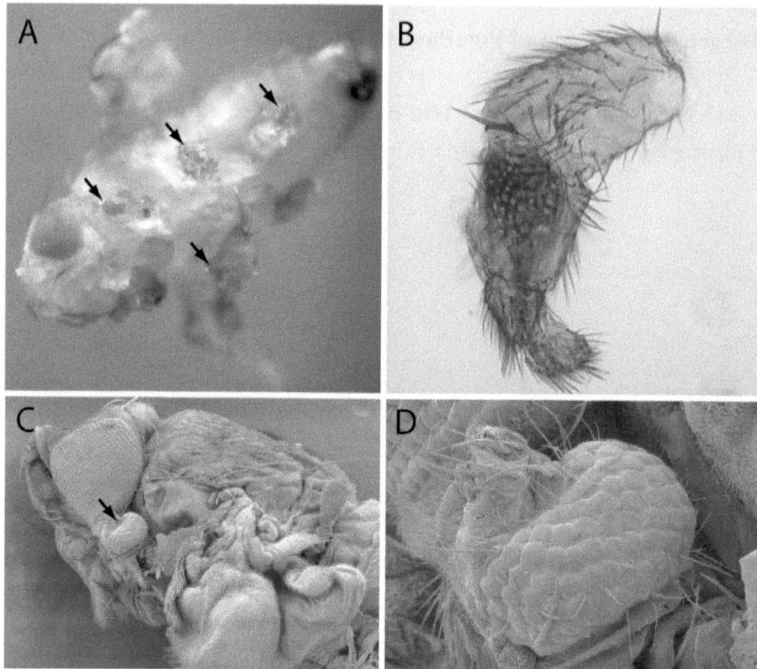

Figure 1.9.2 Ectopic eye structures in *Drosophila melanogaster* induced by overexpression of the bivalvian *PmaPax6*. (**A, B**) Micrograph of ectopic eyes. (**C, D**) Scanning electron micrograph of ectopic eyes. (**A**) Overview of an adult fly with several ectopic eyes on the legs. (**B**) Dissected leg with a large outgrowth of eye tissue. (**C**) Overview of the fly. (**D**) Higher magnification of C. The ectopic eye has an array of hexagonal ommatidia and interommatidial bristles. Arrows point to ectopic eyes.

2. Pecten maximus Six1/2 (PmaSix1/2)

2.1 Isolation of the PmaSix1/2 full-length cDNA

In search for *Pecten maximus Six* genes a PCR approach was carried out using degenerated primers corresponding to highly conserved amino acid sequences found in all Six subclasses. The forward primer corresponds to the conserved N-terminal region QVACVC of the Six domain, whereas a reverse primer was designed corresponding to the C-terminal region NWFKNRRQR of the Six homeodomain. PCR was conducted on eye-specific cDNA templates.

A 498 bp long fragment was isolated, designated *Pma*SixPCR1 (Figure 2.1.1), spanning nearly the entire six domain and homeodomain. A BLAST search against the GenBank database using the isolated fragment showed high sequence homology to the Six1/2 subclass. To obtain the full sequence information of *PmaSix1/2* cDNA the obtained fragment was extended by RACE using specific nested primers resulting in two fragments designated *Pma*Six3'RACE and PmaSix5'RACE (Figure 2.1.1).

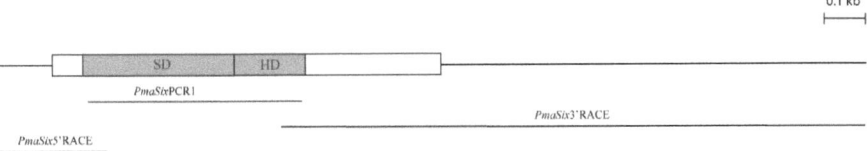

Figure 2.1.1 Schematic of the *PmaSix1/2* full length cDNA. The open reading frame is boxed and the Six domain and Homeodomain are indicated by grey boxes. The PCR clone, 3' RACE clone and the 5' RACE clone are shown underneath.

2.2 Nucleotide and deduced amino acid sequence of *PmaSix1/2*

The full length *PmaSix1/2* cDNA is 2082 bp long with a putative open reading frame of 924 bp encoding a 308 amino acid long protein (Figure 2.2.1). A putative initiator methionine is located 23 codons upstream of the encoded six domain. The six domain has a length of 115 amino acids followed by a 60 amino acid long six homeodomain. The C-terminal region which follows the six homeodomain has a length of 111 amino acids rich in serine (11.8%), proline (9.7%) and threonine (7.3%). The stop codon is found at position 1048, 76 codons upstream of the encoded homeodomain.

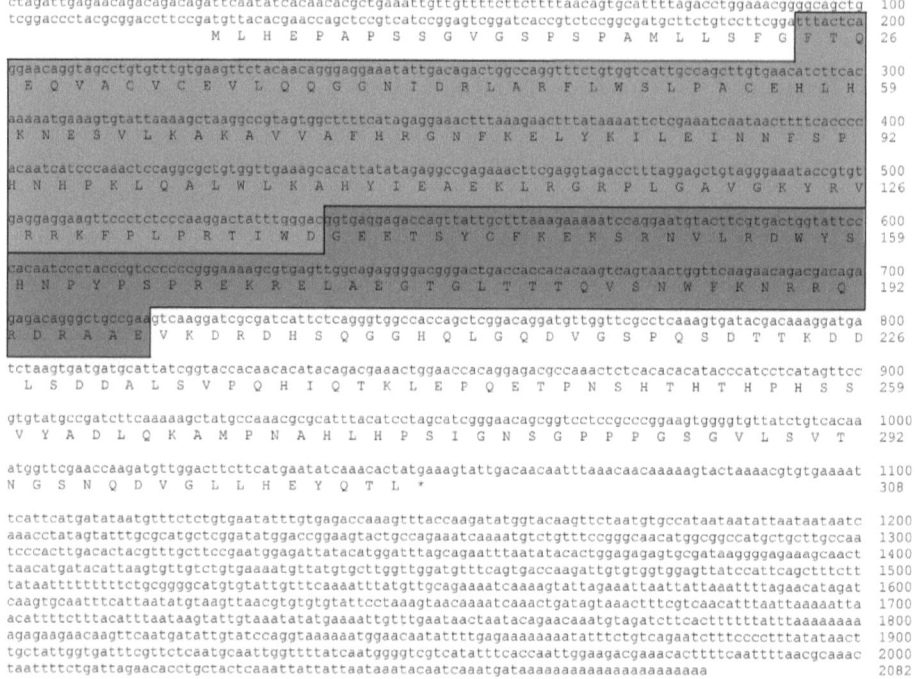

Figure 2.2.1: Nucleotide and deduced amino acid sequence of *PmaSix1/2* full-length cDNA. The six domain is boxed in red, the six homeodomain is boxed in green.

2.3 The six domain

Within the six domain PmaSix1/2 shows high sequence homology to the corresponding region of AnSix1/2 (97% sequence identity; Figure 2.3.1). Only four amino acid substitutions are found compared to AnSix1/2 (position 45, 64, 68 and 85). Beside the high sequence homology to AnSix1/2, the six domain of PmaSix1/2 shows highest sequence identity to the six domain of the lophotrochozoan polychaete *Platynereis* Six1/2 homolog (96%). Much lower sequence identity is found to the cnidarian *Cladonema* Six1/2 (76%) and the planarian *Dugesia* Six1/2 protein (82%).

The six domain of PmaSix1/2 diverges only at one site compared to the majority of Six1/2 homologs (position 64).

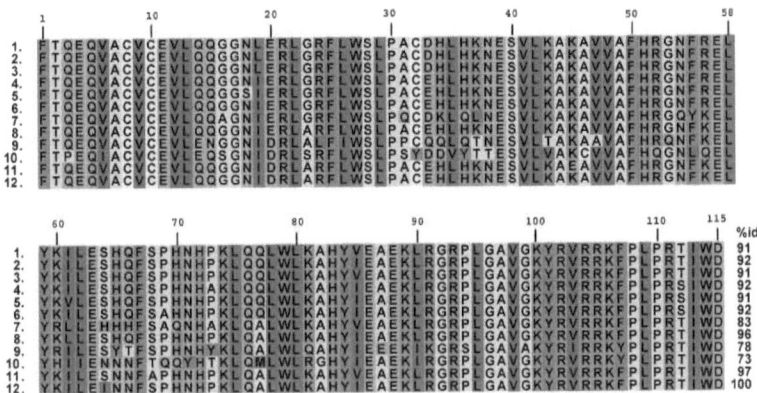

2.4 The six homeodomain

As typical for Six1/2 homeodomains, PmaSix1/2 exhibit the diagnostic tetrapeptide ETSY at the N-terminal part of the homeodomain (Figure 2.4.1). The PmaSix1/2 homeodomain shows highest sequence similarity to the *Drosophila* sine oculis homeodomain (95%; Figure 2.4.1). In general, high sequence identities are obvious between the six homeodomain of PmaSix1/2 and that of other Six1/2 homologs, often exceeding 90%. Lower sequence homology is found for the six homeodomain of *Dugesia* Six1/2. However, the recognition helix are identical in all Six1/2 homologs (Figure 2.4.1).

```
1. SIX1 Homo sapiens
2. Six1 Danio rerio
3. Six1 Gallus gallus
4. SIX2 Human
5. Six2 Danio rerio
6. Six2 Gallus gallus
7. Sine oculis Drosophila melanogaster
8. Six1/2 Platynereis dumerilii
9. Six1/2 Dugesia japonica
10. Six1/2 Cladonema radiata
11. Six1/2 Arca noae
12. Six1/2 Pecten maximus
```

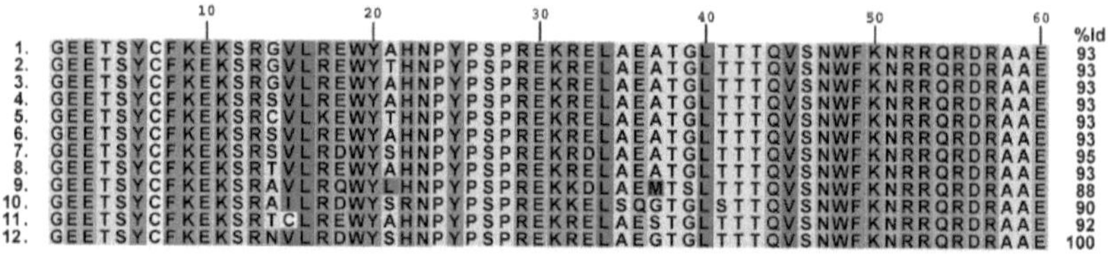

	10	20	30	40	50	60	%Id
1.	GEETSYCFKEKSRGVLREWYAHNPYPSPREKRELAEATGLTTTQVSNWFKNRRQRDRAAE	93
2.	GEETSYCFKEKSRGVLREWYTHNPYPSPREKRELAEATGLTTTQVSNWFKNRRQRDRAAE	93
3.	GEETSYCFKEKSRGVLREWYAHNPYPSPREKRELAEATGLTTTQVSNWFKNRRQRDRAAE	93
4.	GEETSYCFKEKSRSVLREWYAHNPYPSPREKRELAEATGLTTTQVSNWFKNRRQRDRAAE	93
5.	GEETSYCFKEKSRCVLREWYTHNPYPSPREKRELAEATGLTTTQVSNWFKNRRQRDRAAE	93
6.	GEETSYCFKEKSRSVLREWYAHNPYPSPREKRELAEATGLTTTQVSNWFKNRRQRDRAAE	93
7.	GEETSYCFKEKSRSVLRDWYSHNPYPSPREKRDLAEATGLTTTQVSNWFKNRRQRDRAAE	95
8.	GEETSYCFKEKSRTVLREWYAHNPYPSPREKRELAEATGLTTTQVSNWFKNRRQRDRAAE	93
9.	GEETSYCFKEKSRAYLRQWYLHNPYPSPREKKDLAEMTSLTTTQVSNWFKNRRQRDRAAE	88
10.	GEETSYCFKEKSRAILRDWYSRNPYPSPREKKELSQGTGLSTTQVSNWFKNRRQRDRAAE	90
11.	GEETSYCFKEKSRTCLREWYAHNPYPSPREKRELAESTGLTTTQVSNWFKNRRQRDRAAE	92
12.	GEETSYCFKEKSRNVLRDWYSHNPYPSPREKRELAEGTGLTTTQVSNWFKNRRQRDRAAE	100

Figure 2.4.1 Sequence comparison of the six homeodomains of different species. The percentage of sequence identity to the respective AnSix1/2 amino acid sequence are indicated at the end of each line.

2.5 Phylogenetic analysis of *A. noae* and *P. maximus* Six1/2

A phylogenetic tree was constructed on the basis of the full six domain and homeodomain of various Six homologs using the neighbour joining method (Figure 2.5.2). The phylogenetic analysis confirms the classification of the *Arca* and *Pecten Six* genes into the Six1/2 subfamily. Moreover *AnSix1/2* and *PmaSix1/2* clusters together, indicating the close phylogenetic relationship between *Arca* and *Pecten*.

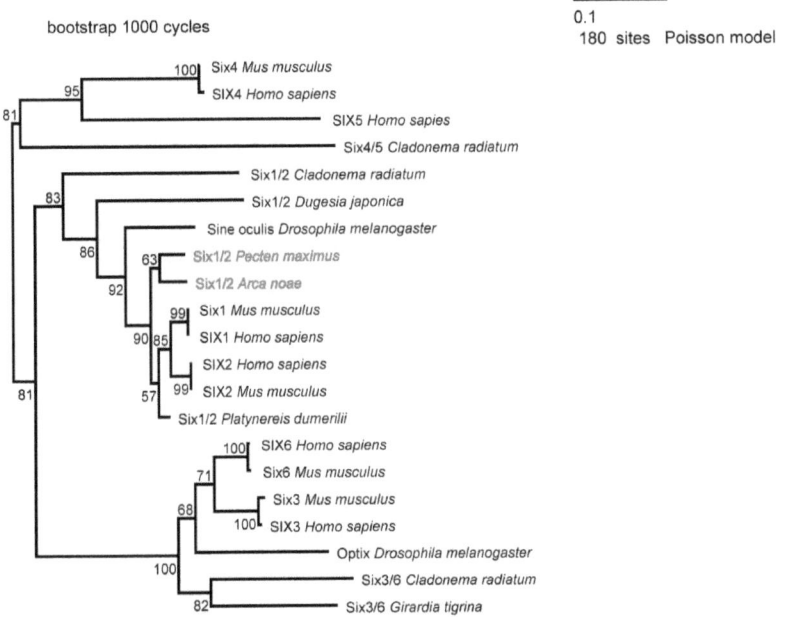

Figure 2.5.1 Phylogenetic analysis of *Arca* and *Pecten* Six proteins (highlighted in red) confirms their classification into the Six1/2 subfamily. The tree was generated by the neighbour joining method using 180 sites from the full six domain and homeodomain. Bootstrap percentage values (1000 replicates) are indicated over the corresponding nodes.

2.6 Real-time PCR expression analysis of *PmaSix1/2*

To get a general overview of *PmaSix1/2* expression, a real-time PCR expression analysis in various tissues was performed. Messenger RNA was isolated from eyes, mantle tissue, muscle tissue, ovary and gill tissue and reverse transcribed using random primers to generate cDNA templates. Specific primers for real-time PCR were designed for an amplicon size of 142 bp within the Six domain. Three independent real-time PCR expression analysis were carried out on cDNA templates generated from mRNA isolated from three different individuals. The *Pecten*-specific housekeeping *Ef1α* was used as a reference for normalization and to compensate for variations in the quantity and quality of the cDNA preparations.

Real-time PCR expression data show only little *PmaSix1/2* expression in the eye relative to other examined tissues (Figure 2.6.1). Also in the ovaries, expression levels are low, comparable to expression levels found in the eye. Real-time PCR analysis suggests slightly more *PmaSix1/2* expression in the mantle tissue than in the eye and ovary tissue. Significant higher expression is found in the gill. However, no *PmaSix1/2* expression was found in any of the examined tissues by *in situ* hybridization, suggesting that *PmaSix1/2* is expressed below the threshold levels to detect transcripts by *in situ* hybridization. Consistent with low expression levels is the observation of high C_T-values indicated by the real-time PCR amplification plot (Data not shown).

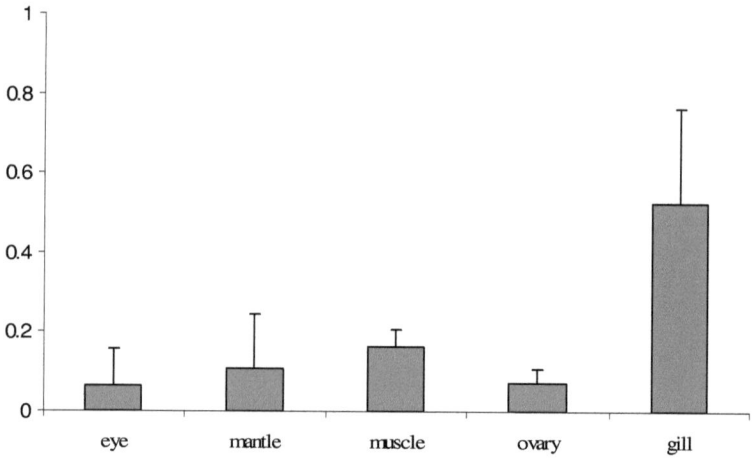

Figure 2.6.1 *PmaSix1/2* expression analysis in various tissues of *Pecten*. Graphs display the relative values normalized to elongation factor expression levels.

2.7 Expression analysis of *PmaSix1/2* in *Pecten* larvae by whole mount *in situ* hybridization

No *PmaSix1/2* expression was detected in the eye nor in any other tissue of adult animals by *in situ* hybridization. In *Pecten* veliger larvae (stage 21 to 24), however, expression was found in three distinct areas (Figure 2.7.1). Because the staining was rather weak and faded when examined by higher magnifications under the light microscope, it was

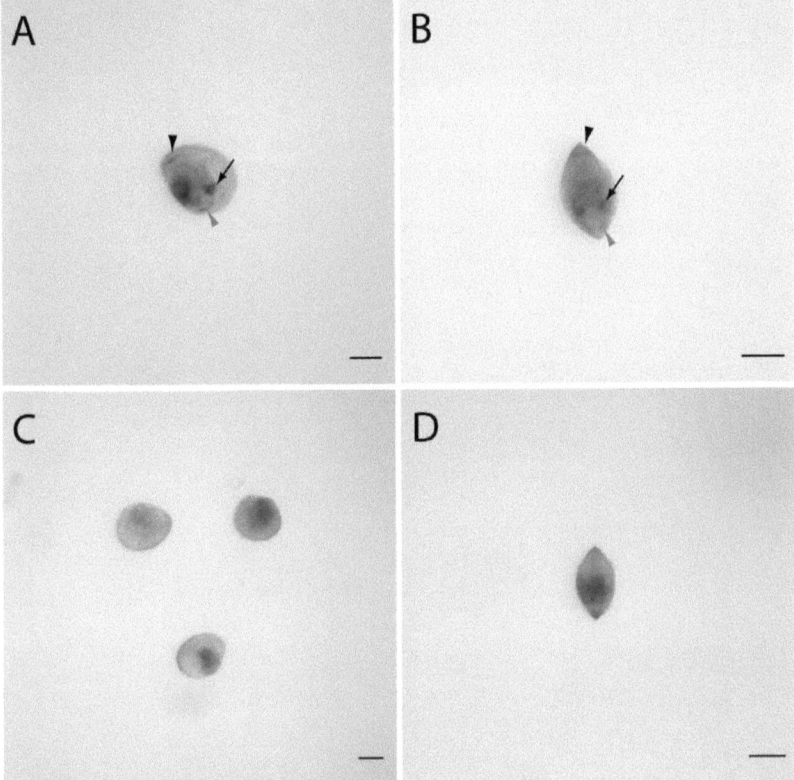

Figure 2.7.1 *In situ* hybridization on veliger larvae (stage 24) using DIG-labeled probes for *PmaSix1/2*. (**A**) and (**C**) Lateral view of *Pecten* veligers. (**B**) and (**D**) Ventral views of *Pecten* veligers. (**A**) Lateral view of veliger larva indicating staining in the putative gill primordia (arrow), the putative posterior adductor muscle (red arrowhead) and the putative anterior adductor muscle (black arrowhead). (**B**) Ventral view of veliger larva shows bilateral *PmaSix1/2* expression in the putative gill primordia (black arrow) and the putative posterior (red arrowhead) and anterior (black arrowhead) adductor muscles. (**C**) and (**D**) Negative control using a sense probe for *PmaSix1/2*. Scale lines: 100µm.

difficult to identify superimposed tissues. Therefore, it was not possible to assign the staining to the respective tissues with absolute certainty. However, the anatomical locations of the stainings suggests expression in the anterior and posterior adductor muscles (Fig 2.7.1A and B).

Moreover, expression was found in a region just anterior to the posterior adductor muscles, which most likely correspond to the gill primordia (Fig. 2.7.1A and B)

3. Opsin genes in Pecten maximus

The scallop mirror eye is unique in that it contains two layers of photoreceptor cells: a proximal retina with rhabdomeric photoreceptor cells and a distal retina with ciliary photoreceptor cells. In the pacific scallop *Patinopecten yessoensis*, two distinct opsin genes have been isolated (Kojima et al., 1997). One of them, *Scop2* (or Go-coupled rhodopsin) was found to be expressed in the ciliary retinal layer. The second, *Scop1* (or Gq-coupled rhodopsin), was suggested to be expressed in the rhabdomeric photoreceptor cells since an antibody against the squid Gq-coupled rhodopsin cross-reacted to the scallop proximal retina.

3.1 Isolation of two *Pecten* opsin genes (*PmaGqOpsin* and *PmaOpsinX*)

In search for opsin genes in *Pecten maximus* a low-stringency PCR approach was performed using degenerated primers as already illustrated for the isolation of *AnOpsinX*. From eye specific cDNA templates it was possible to isolate two 474bp long fragments, designated *Pma*GqOpsPCR1 and *Pma*OpsPCR1 (Figure 3.1.1A and B). When compared to the databases (GenBank), the *Pma*GqOpsPCR1 fragment exhibited 55% amino acid identity to the scallop *Patinopecten yessoensis* Gq-coupled (rhabdomeric) rhodopsin, 52% identity to the octopus *Pareledone turqueti* rhodopsin and 49% identity to the rhabdomeric opsin of *Platynereis dumerilii*. In contrast, the latter fragment showed only little sequence identity when compared to the databases (GenBank). Highest amino acid sequence identity was found to a yet unspecified mosquito opsin genes of *Aedes aegypti* (42%) and *Anopheles gambiae*

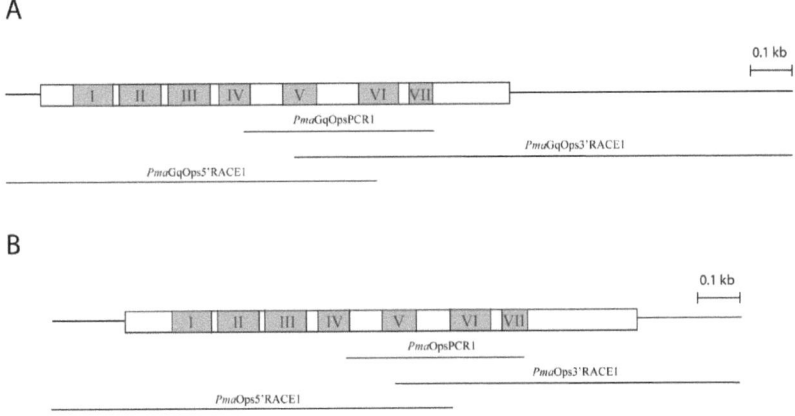

Figure 3.1.1 Schematic of *PmaGqOpsin* (**A**) and *PmaOpsinX* (**B**) full length cDNA. The open reading frames are boxed and the transmembrane domains are indicated by grey boxes serially numbered. The PCR clone, 3' RACE and 5' RACE clones are shown underneath.

(38%).

The opsin fragments were subsequently extended by RACE using specific primers yielding two clones for each opsin gene, designated *Pma*GqOps5'RACE1 / *Pma*GqOps3'RACE1 and *Pma*Ops5'RACE1 / *Pma*Ops3'RACE1, respectively (Figure 3.1.1A and B).

3.2 Nucleotide and deduced amino acid sequence of *PmaGqOpsin* full-length cDNA

The full length *PmaGqOpsin* cDNA has a length of 1978 bp encoding for a open reading frame of 372 amino acids (Figure 3.2.1). A putative initiation site is found 26 codons upstream of the encoded helix I. The termination codon is found at position 1202, 64 codons upstream of helix VII. The 5'-untranslated region has a putative length of 85 bp. In contrast, the 3'-untranslated region is much longer with a putative length of 777 bp. By multiple alignments of opsins and comparison to the secondary structure of bovine rhodopsin (Palczewski et al., 2000) it was possible to identify seven transmembrane helices. Moreover,

Figure 3.2.1 Nucleotide and deduced amino acid sequence of *PmaGqOpsin* cDNA. Boxed amino acid sequences (red) correspond to the transmembrane domains, numbered form I to VII. The highly conserved lysine in the seventh transmembrane domain which is characteristic for opsin proteins is underlined.

the lysine residue specific for opsin proteins was found in helix VII of PmaGqOpsin (Figure 3.2.1)

3.3 Nucleotide and deduced amino acid sequence of *PmaOpsinX* full-length cDNA

The full-length *PmaOpsinX* has a length of 1612bp encoding for a 397 amino acid long protein (Figure 3.3.1). A putative initiator methionine is found 37 codons upstream of helix I and the termination codon is found 86 codons (position 1362) upstream of helix VII. The 5'-untranslated region has a putative length of 170bp, whereas the 3'-untranslated region has a putative length of 251bp.

By multiple alignments and comparison to the two-dimensional model of bovine rhodopsin (Palczewski et al., 2000), seven putative transmembrane helices have been identified. In addition, the opsin specific lysine, which is essential for the binding of the chromophore, is found at the appropriate position (K302) in helix VII (Figure 3.3.1).

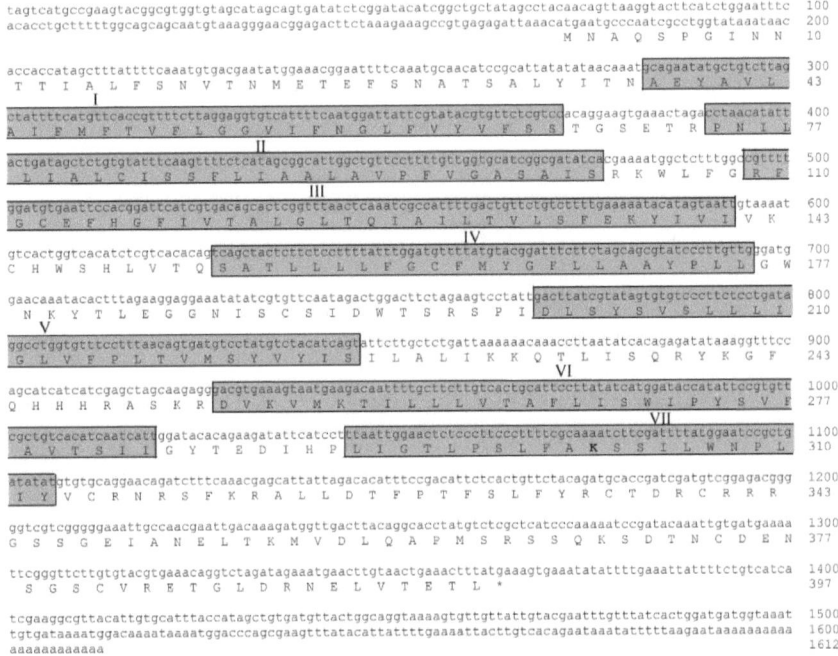

Figure 3.3.1 Nucleotide and deduced amino acid sequence of *PmaGqOpsin* full-length cDNA. Boxed sequences (red) correspond to the seven transmembrane domains, numbered from I to VII. The highly conserved lysine residue in the seventh transmembrane helix is underlined.

3.4 Structural analysis of PmaGqOpsin

Several hydrophobic segments are suggested by the hydropathy plot of PmaGqOpsin (Figure 3.4.1). Of the seven putative transmembrane helices commonly found in opsins, six hydrophobic segments are obvious from the plot. Since the seventh helix is quite polar in the middle, with a lysine and a serine residue, it does not show up very well.

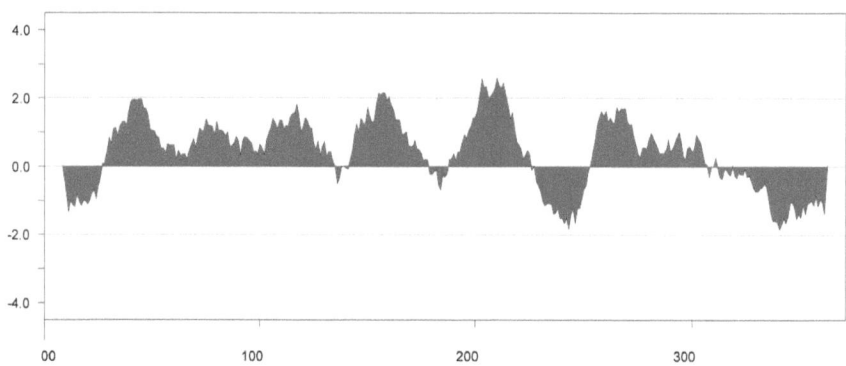

Figure 3.4.1 Hydropathy blot of PmaGqOpsin. Six of seven transmembrane domains of PmaGqOpsin are apparent in the hydropathy blot.

A two-dimensional model of PmaGqOpsin was generated (Figure 3.4.2) by multiple sequence alignment and by comparison to the two-dimensional model of bovine rhodopsin (Palczewski et al., 2000). Typically, the highest degree of variation is found in the intracellular loops of PmaGqOpsin, in particular the third cytoplasmic loop (C-III; Figure 3.4.2).

The highly conserved asparagine residue (N55 in bovine rhodopsin) found in the first helix of most GPCRs is also in helix I at position 47 of PmaGqOpsin.

Two lysine residues (K58 and K61) are found in the first cytoplasmic loop of PmaGqOpsin (Figure 3.4.2), consistent with the finding that the first cytoplasmic loop of GPCRs generally have several positively charged amino acids which may be important for the proper insertion into the membrane (Hartmann et al., 1989). At the N-terminal end of helix III, a highly conserved cysteine among GPCRs is found at position 103 in PmaGqOpsin which is likely to be engaged in a disulfide bond with C181.

At the C-terminal end of helix III, PmaGqOpsin has a highly conserved motif, E(D)RY, which is suggested to regulate the interaction of GPCRs with their corresponding G-protein (Figure 3.4.2). Two out of three amino acid residues are conserved in PmaGqOpins (E127 and R128), whereas the third is substituted by a cysteine (C129).

A tyrosine residue at position 106 serves as a counterion of the protonated Schiff base in PmaGqOpsin (Figure 3.4.2).

The third cytoplasmic loop, C-III, is highly variable in GPCRs and is highly critical for G-protein binding. In PmaGqOpsin, C-III shows highest amino acid identity to Gq-coupled rhodopsins (rhabdomeric opsins) with the highest homology to scallop (Scop1) and squid rhodopsin.

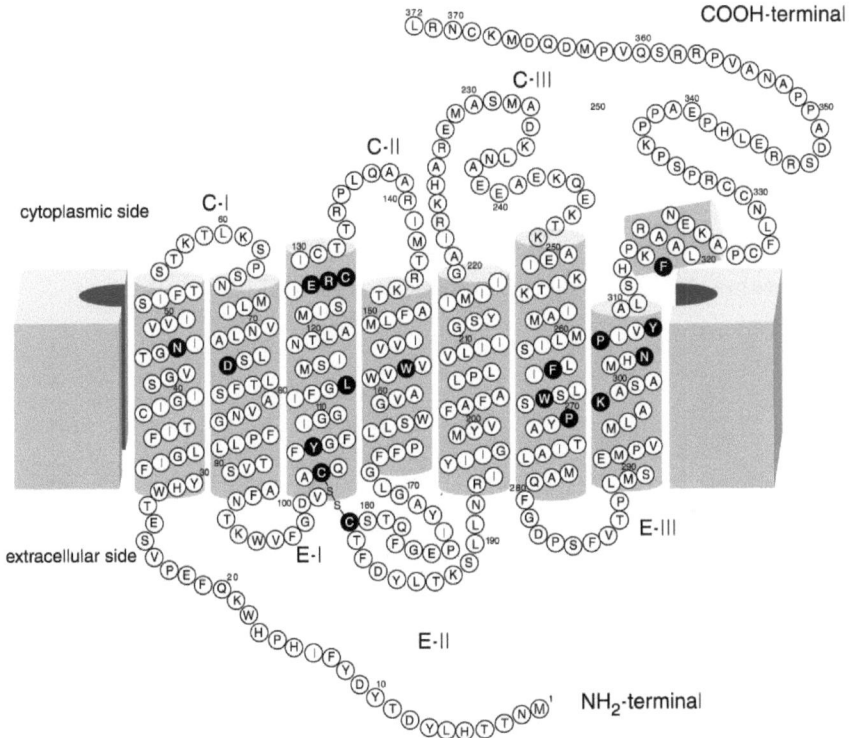

Figure 3.4.2 Two-dimensional model of PmaGqOpsin (modified after Palczewski et al., 2000). Some of the key residues which are conserved in opsins are shown in filled circles.

Four highly conserved amino acids, F261, W265, Y268 and A269 were found to surround the β-ionone ring of retinol in bovine rhodopsin (Palczewski et al., 2000) In PmaGqOpsin, these amino acids are located at the corresponding positions F261, W265, Y268 and A269 of helix VI (Figure 3.4.2).

The critical lysine residue of helix VII is found at position K299 in PmaGqOpsin (Figure 3.4.2). Moreover, the highly conserved NPXXY motif is conserved in PmaGqOpsin, corresponding to N305, P306, I307, V308 and Y309. Adjacent to helix VII, there is a putative cytoplasmic α-helix, spanning the region from P314 to C326 in PmaGqOpsin.

A stretch of three amino acids in the C-terminus is highly indicative of the ciliary-opsin and rhabdomeric-opsin families (Arendt et al., 2004). In PmaGqOpsin, a HPK-motif (H313, P314 and H315) was detected which is shared among rhabdomeric opsins (Figure 3.4.2).

3.5 Structural analysis of PmaOpsinX

The hydropathy blot confirms the existence of several hydrophobic elements in PmaOpsinX (Figure 3.5.1). As typical for GPCRs, six hydrophobic segments of seven putative transmembrane helices are obvious, whereas the seventh, which is quite polar in the middle (a lysine and two serines in PmaOpsinX) showed up less evidently.

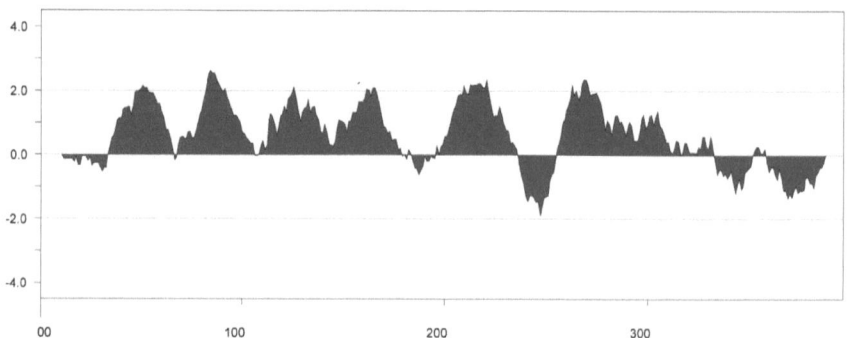

Figure 3.5.1 Hydropathy plot of PmaOpsinX. Six of seven putative transmembrane domains are obvious in the hydropathy plot.

A two-dimensional model of PmaOpsinX was generated (Figure 3.5.2) by multiple alignments and by comparison to the two-dimensional model of bovine rhodopsin (Palczewski et al., 2000).

As typical for opsins, the first transmembrane domain of PmaOpsinX has a highly conserved asparagine at position N78 (Figure 3.5.2). Generally, GPCRs have several positively charged amino acids in the first cytoplasmic loop and which are suggested to be important for the correct insertion of the protein into the membrane (Hartmann et al., 1989). Deviating from the general rule, PmaOpsinX has only one positively charged amino acid

(R73) in the first cytoplasmic loop. A characteristic cysteine at the N-terminal end of helix III, common to all GPCRs, is also found at position S112 of PmaOpsinX and is most likely engaged in a disulfide bond with C189. In contrast to rhabdomeric opsins and ciliary opsins which usually have a glutamate or a tyrosine as a counterion, PmaOpsinX has a histidine at the corresponding position (H113).

The proline residue (P267 in bovine rhodopsin) of helix VI, which was found to be highly conserved among GPCRs and leads to a strong bending of the helix is found at the corresponding position 273 in PmaOpsinX (Figure 3.5.2). Moreover, the conserved amino acid residues of helix VI, which were shown to surround the β-ionone ring of retinol in bovine rhodopsin, are also found in PmaOpsinX (F268, W271 and Y274) with the only exception of S265 which replaces the alanine residue commonly found in GPCRs.

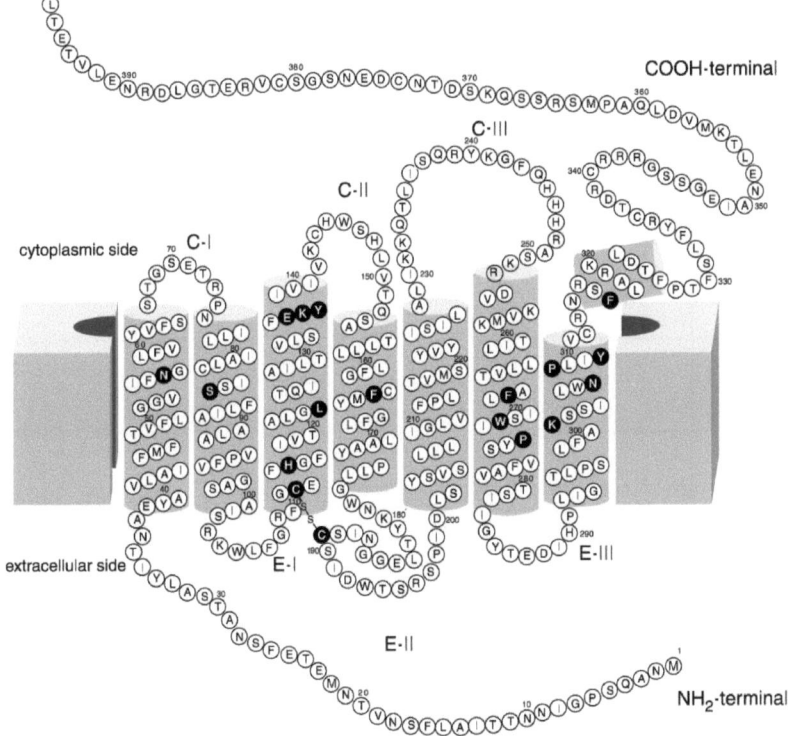

Figure 3.5.2 Two-dimensional model of PmaOpsinX. Modified after(Palczewski et al., 2000). Some of the key residues which are conserved in GPCR are shown in filled circles.

The highly conserved and functionally important lysine of helix VII is found at position K302 in PmaOpsinX. Furthermore, the NPXXY motif typical for opsins is also found in PmaOpsinX, corresponding to N308, P309, L310, I311 and Y312 (Figure 3.5.2).
A putative cytoplasmic α-helix, spanning the region from R317 to T329 is found just adjacent to helix VII.

3.6 Phylogenetic analysis of *Arca* and *Pecten* opsin genes

The construction of a phylogenetic tree by the neighbour joining method revealed that the *Pecten* Gq-coupled rhodopsin gene clearly clusters within the rhabdomeric opsin subfamily (Figure 3.6.2A). By the neighbour joining method it was not possible to assign the *Arca* opsin gene *AnOpsinX* and the *Pecten* opsin gene *PmaOpsinX* to any of the known opsin subfamilies with statistical significance. The main reason is that the NJ method is based on the Distance Method which generally leads to the loss of information.

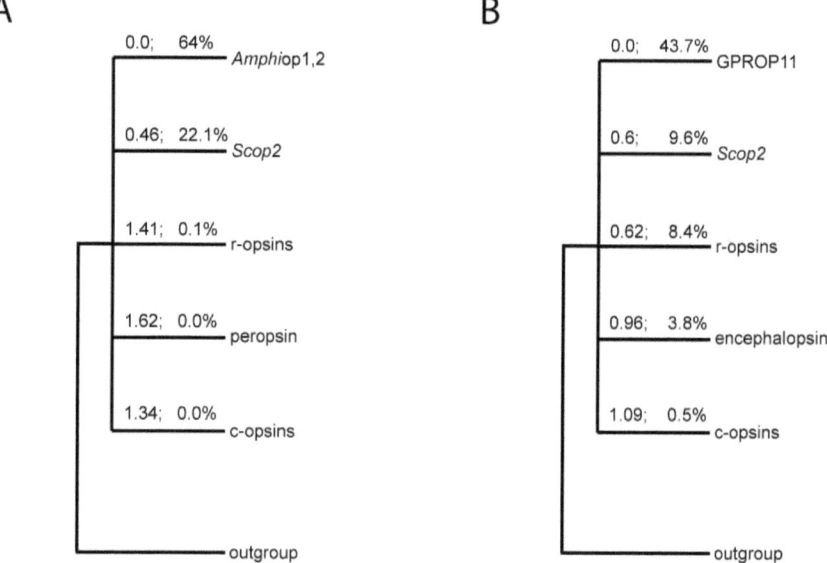

Figure 3.6.1 Local bootstrap probabilities estimated for alternative assignments of *AnOpsinX* (**A**) and *PmaOpsinX* (**B**), respectively, to various opsin subfamilies. Differences of log likelihood (dlkl) and percentage of local bootstrap probabilities are indicated over the branches. c-opsin: ciliary opsins, r-opsins: rhabdomeric

Therefore, a phylogenetic tree was generated (Figure 3.6.2) by the maximum likelihood method and estimated the local bootstrap probabilities (LBP) for different tree topologies by using the MOLPHY program (Adachi and Hasegawa, 1996).

Phylogenetic analysis indicates that the *Arca* opsin gene (*AnOpsinX*) most likely (dlkl = 0.0; distance of log likelihood) clusters with the Go-coupled opsins (*Amphi*op1 and *Amphi*op2) (Koyanagi et al., 2002) of Amphioxus (Figure 3.6.1A). However, it can not be excluded that *AnOpsinX* clusters to the Go-coupled opsin *Scop2* of the pacific scallop *Patinopecten* (Kojima et al., 1997) (dlkl = 0.46; Figure 3.6.1A). However, the possibility that *AnOpsinX* clusters to c-opsin, r-opsin or peropsin subfamilies can be excluded (dlkl > 1.0; Figure 3.6.1A). With a LBP of 13.8% (Figure 3.6.1A) the possibility can not be excluded that *AnOpsinX* does not cluster to any of the examined opsin subfamilies and therefore may constitute a new subfamily.

The *Pecten* opsin gene, *PmaOpsinX*, most likely clusters to the yet uncharacterized mosquito opsin gene *gropl1* (dlkl = 43.7%; Figure 3.6.1B). Weak support is found for the possibility that *PmaOpsinX* clusters to *Scop2* (dlkl = 0.6), rhabdomeric opsins (dlkl = 0.62) or encephalopsin (dlkl = 0.96) (Figure 3.6.1B). The possibility that *PmaOpsinX* clusters to ciliary opsin subfamilies can be excluded (dlkl > 1.0; Figure 3.6.1B). With a LBP of 34%, the possibility can not be excluded that *PmaOpsinX* does not cluster to any of the examined opsin subfamilies and may, therefore, constitute a new opsin subfamily (Figure 3.6.1B).

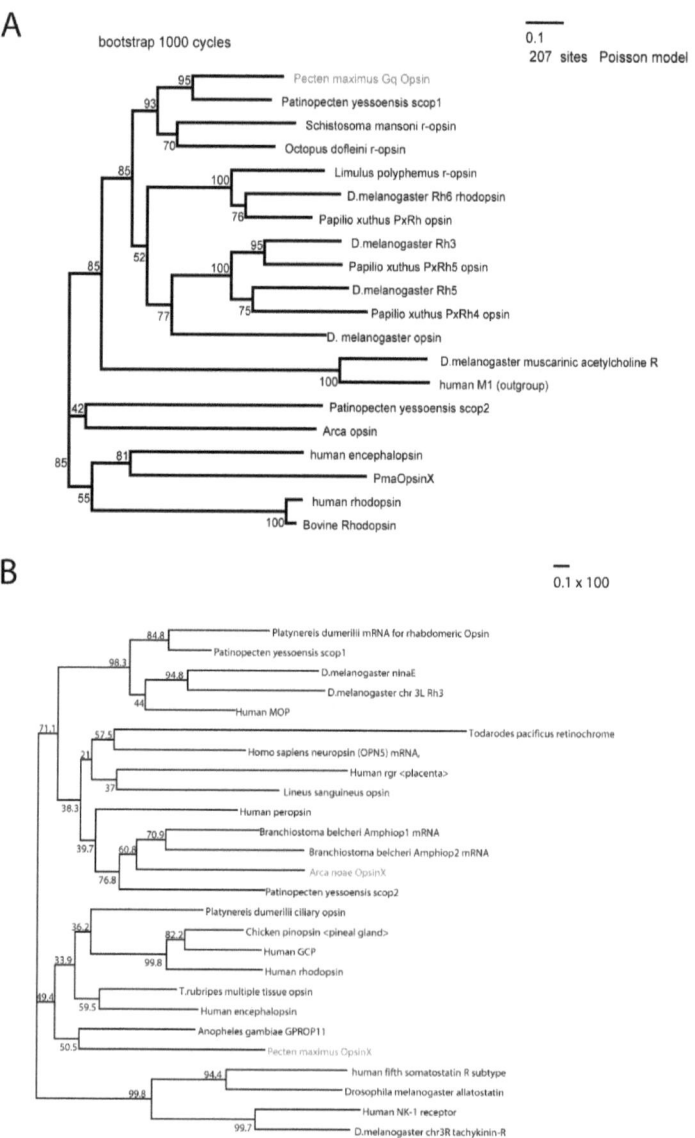

Figure 3.6.2 Phylogenetic trees of visual pigments. **(A)** Phylogenetic analysis of *PmaGqOpsin* (red) by the NJ method demonstrates its classification as a rhabdomeric opsin. Bootstrap percentage values (1000 replicates) are indicated over the corresponding nodes. **(B)** Phylogenetic analysis of *AnOpsinX* and *PmaOpsinX* (red) by the ML method. Local bootstrap percentage values are indicated over the corresponding nodes.

3.7 Real-time PCR expression analysis of *PmaGqOpsin* and *PmaOpsinX*

Two opsin genes were isolated from *Pecten*. By sequence comparison and phylogenetic analysis it was possible to unambiguously identify *PmaGqOpsin* as a Gq-coupled rhodopsin most closely related to the pacific scallop *Scop1* gene. Due to the fact that *PmaOpsinX* showed no obvious homology to any of the known eye-specific opsin genes, the possibility that *PmaOpsinX* may have a non-visual function and might also be expressed outside of the eye was considered. Therefore, to get a general overview of opsin expression in various tissues, a real-time PCR expression analysis was performed. Eyes, mantle tissue, adductor muscle tissue, ovary tissue and gill tissue were used to isolate mRNA and to generate cDNA. Expression analysis was done three times on independent cDNA templates generated from three independent individuals. The *Pecten* Elongation factor *Pmaef1α* was used to compensate for variations in the quality and quantity of cDNA preparations and for normalization. As expected, *PmaGqOpsin* was found to be expressed exclusively in the eye (Figure 3.7.1A). In contrast, *PmaOpsinX* showed expression in all examined tissues, with the strongest expression found in the eye and the gill (Figure 3.7.1B). However, the high C_T-values indicated by the real-time PCR amplification plot and the failure to detect any *PmaOpsinX* expression by *in situ* hybridization suggests low expression levels, probably below the threshold levels needed to detect *in situ* hybridization signals.

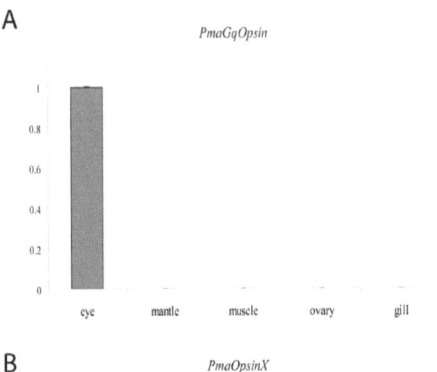
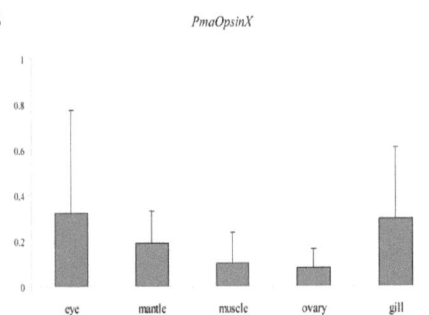

Figure 3.7.1 Expression analysis of Pecten opsins by real-time PCR. (A) *PmGqOpsin* expression; (B) *PmOpsinX* expression. Graphs display relative values normalized to elongation factor expression level.

3.8 Expression analysis of *PmaGqOpsin* in *Pecten* eyes by *in situ* hybridization

Consistent with our data from real-time PCR which suggest high expression in the mirror eyes of *Pecten*, strong staining signals were detected on cryo-sections of *Pecten* eyes by *in situ* hybridization (Figure 3.8.1A)

Expression of *PmaGqOpsin* was found to be limited to the proximal rhabdomeric cell layer (Figure, 3.8.1A)). This is consistent with the recent finding that a rhabdomeric specific G_q-type is expressed in the rhabdomeric cell layer of the pacific scallop *Patinopecten yessoensis* (Kojima et al., 1997).

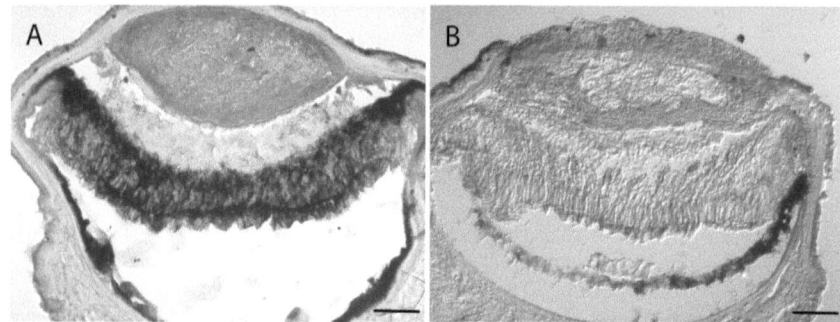

Figure 3.8.1 *In situ* hybridization on cryo-sections of the *Pecten* mirror eye using a DIG labeled probe for *PmaGqOpsin*. (**A**) Expression of *PmaGqOpsin* is restricted to the proximal rhabdomeric photoreceptor cell layer. (**B**) Control experiment using a DIG labeled sense probe for *PmaGqOpsin*. Bar = 0.1mm.

V. Discussion

Understanding how eyes evolved requires analysis of the key regulatory genes involved in eye development. While morphological comparisons of eye anatomy led to the view that various eye-types arosed independently during evolution (Salvini-Plawen and Mayr, 1977), the finding that *Pax6* has a conserved functions during eye development in various eumetazoans suggests that animal eyes evolved from a common, simple proto-type eye (Gehring and Ikeo, 1999).

In order to gain further evidence for this idea, the investigations were focused on two bivalvian species, *Arca noae* and *Pecten maximus*, each representing a different eye-type, making them suitable animals to test the hypothesis of monophyletic eye evolution.

To introduce new animal models into a laboratory is often associated with several constraints. Due to the lack of functional assays, such as transgenic methods or appropriate cell lines, our work was restricted to more descriptive research. In addition, a bottleneck of our investigation was the inability to obtain early developmental stages.

Spawning and gametogenesis are seasonal in scallops and *Arca* (Berg and Strand, 2001; Peharda et al., 2006). There are one or two spawning peaks each year, depending on the geographical location of the population making the collection of embryos and larvae difficult. In general, artificial fertilization of *Pecten maximus* is possible. However, the animals have to be conditioned for spawning, since passage of the oocytes through the oviduct is necessary to reactivate meiosis (Robert and Gérard, 1999) and therefore, in contrast to many other sea animals, it is not possible to get ripe eggs by stripping. Well-elaborated facilities and excellent sea-water conditions together with appropriate nutrition supply are essential to conduct successful artificial fertilization (personal communication, P. Miner). Because of the lack of appropriate facilities and the lack of technical know-how, it was beyond our power to carry out artificial fertilization to study development. Thanks to IFREMER in Brest (France), it was finally possible to get access to at least some larval stages at the end of this PhD. Unfortunately, our investigations were limited to only a few developmental stages (day 16 to day 27 after fertilization). Nevertheless, we succeeded to collect *Pecten* larvae from developmental stages where the larval eyes start to appear.

1. The *Arca noae* compound eye

The compound eyes of *Arca noae* differ from those of arthropods in having only one photoreceptor cell per ommatidium instead of eight or nine visual cells. Another striking difference is the usage of ciliary photoreceptor cells, quite contrary to the universal use of rhabdomeric photoreceptor cells in arthropod compound eyes.

In addition, in contrast to compound eyes in arthropods, no real lens structures were found in the compound eyes of *Arca*. This is consistent with the previous finding that the compound eyes of the closely related ark clam *Barbatia cancellaria* have neither a lens nor any other focusing structures (Nilsson, 1994). However, Nilsson (Nilsson, 1994) has not taken into consideration that the numerous small mitochondria, distally located to the photoreceptive element may have a lens-like function. Support for this idea comes from few cases, reported from planarians, where the lenses are of mitochondrial origin (Rhode and Watson, 1991) or build up by numerous small mitochondria (Bedini et al., 1973).

Further investigations of the optics will be needed to ascertain whether the mitochondrial arrangements in *Arca* indeed have a focusing function.

Interestingly enough, Electron micrographs of *Arca* compound eyes show a huge amount of vesicles appearing to bud off from the microvilli that fill the pigment funnel above the photoreceptor cell. Although this feature is also observed in *Barbatia* (Nilsson, 1994), it is much more pronounced in *Arca*. Therefore, it can not be ruled out that the high density of vesicles may play a role as refractive devices in *Arca*.

Another interesting feature of the *Arca* compound eye are the numerous microvilli that fill the pigment funnel above the photoreceptor cell. From the electron micrographs it can not be deduced whether the microvilli are formed by the ciliary photoreceptor cell or by the surrounding pigment cells. Independent of their origin we propose two equal possibilities for their function. Either they may serve as guiding devices for light-rays or they may serve as a second receptive field for light perception. For further studies it will be warranted to search for opsin genes expressed in the microvilli.

Another interesting question is whether the ciliary photoreceptor cells of the non-cerebral eyes of *Arca* are orthologous to the ciliary photoreceptor cells found in the cerebral eyes of vertebrates. Consistent with an orthologous relationship between non-cerebral and cerebral ciliary photoreceptor cells is the finding that ciliary photoreceptor cells of *Pecten* hyperpolarize in response to light (Gorman and McReynolds, 1969). Furthermore, cGMP is used as a second messenger system in both, ciliary photoreceptors of *Pecten* and rod and cones of vertebrates.

Unfortunately, several attempts to investigate the physiological response of visual excitations in *Arca* compound eyes by electroretinogramm and patch clamp failed due to technical problems.

Further investigations on the molecular actors of phototransduction and on the physiological responses induced by photostimulation will be important to shed light on the relationship between non-cerebral eyes of bivalves and cerebral ciliary photoreceptors of vertebrates.

2. The cloning and expression of AnPax6 and PmaPax6

2.1 Sequence conservation of *AnPax6* and *PmaPax6*

Pax6 homologs from two bivalvian species, *Arca noae* and *Pecten maximus* were isolated. The two DNA-binding domains, the paired domain and the paired-like homeodomain, share high sequence homology to all known *Pax6* genes. Importantly, all Pax6-specific amino acids of the paired domain and the homeodomain are conserved in AnPax6 and PmaPax6, with the only exception of a cysteine residue at the N-terminal end of the PmaPax6 paired domain which is usually occupied by an alanine in other Pax6 proteins.

Additionally, sequence homology is found in the eleven amino acid long motif MYDKLGLLNGQ, which has been found in almost all *Pax6* genes and is absent in all other Pax gene families (Loosli et al., 1996). Furthermore, in both, AnPax6 and PmaPax6 we several conserved amino acids (LKRK and KLRNQRR) flanking the homeodomain were found.

Another conserved region is also found at the C-terminal end of Pax6 proteins (Tomarev et al., 1997) which is also found in AnPax6 but lacks in PmaPax6.

AnPax6 and PmaPax6 show an overall amino acid sequence identity of 83%, consistent with the close phylogenetic relationship between *Arca* and *Pecten*. Outside of the bivalvian class, AnPax6 and PmaPax6 show highest overall amino acid sequence identity to the cephalopods *Euprymna scolopes* (72% and 71%, respectively) and *Loligo opalescence* (72% and 71%, respectively) confirming the close phylogenetic relationship between bivalves and cephalopods, both belonging to the phylum of molluscs. Moreover, AnPax6 and PmaPax6 show high overall amino acid sequence identity to the polychaete *Platynereis dumerilii* (65% and 62%, respectively), consistent with the close evolutionary relationship between molluscs and annelids. Although the overall amino acid sequence identity of AnPax6 and PmaPax6 is much lower compared to Pax6 of the nemertine *Lineus sanguineus* (53% and 52%, respectively), high sequence homology is found for the paired domain (95% and 93%) and the homeodomain (97% for both, AnPax6 and PmaPax6).

Phylogenetic analysis using 201 sites from the paired domain, homeodomain and flanking regions of Pax6 confirms the close phylogenetic relationship between molluscs, annelids and nemertines, which all belong to the lophotrochozoan clade.

The high sequence conservation and phylogenetic analysis unambiguously identify *AnPax6* and *PmaPax6* as orthologs of the Pax6 gene.

2.2 Expression of *Pax6* in *Arca* and *Pecten*

Since bivalves continue to grow throughout their lifetime and eyes of different sizes are found along the mantle edge of *Arca* and *Pecten*, it was tempting to assume that new eyes are formed during ongoing growth. However, in contrast to our initial assumption, it was not possible to find any eyes of intermediate developmental stages, beside the general observation that some eyes are obviously smaller than others but do not show any differences in their anatomy compared to larger eyes. Therefore, eye development could not be followed in adult animals.

Nevertheless, we decided to investigate *Pax6* expression in the eyes of adult *Arca* and *Pecten* animals since we assumed that Pax6 may have cell specification functions in the growing eyes. A similar situation, supporting our assumption, is found in the retina of amphibians and fish, which also continue to grow throughout adult life. In these animals, new retinal cells are added from stem cells located in a specialized proliferative zone, the ciliary marginal zone (CMZ) (Johns, 1977; Straznicky and Gaze, 1971; Wetts et al., 1989). Interestingly, in *Xenopus*, the least determined, distalmost CMZ stem cells were shown to express Pax6 (Perron et al., 1998). Pax6 was found to be required for the multipotent state of retinal progenitor cells in the differentiating retina of mice (Marquardt et al., 2001) and remains expressed in the retinal ganglion cells and amacrine cells of the adult retina.

However, no *Pax6* expression was found in the compound eyes of adult *Arca* animals nor in the mirror eyes of adult *Pecten* animals. In both cases, our data from quantitative real-time PCR indicate only very low levels of *Pax6* expression in the eye. Concordantly, no *Pax6* transcripts were detected in the eyes of both species by *in situ* hybridization.

In view of these results, it is possible that *Pax6* does not play a role in the growing process of bivalvian eyes. In favour of this assumption, no expression was found in the developing adult eyes of the polychaete *Platynereis* at the time-point of photoreceptor development, although these eyes exhibit life-long growth and add hundreds of cells to the initial eye primordia (Arendt et al., 2002). Of course the possibility can not be ruled out that *Pax6* may play an important role during the early steps of eye development. Further studies will be necessary to investigate *Pax6* expression and function at animal stages where the peripheral eyes start to develop. So far it is not known when they start to form. We assume, that they likely start to develop in juvenile animals after metamorphosis but early initiation processes may also occur earlier during metamorphosis. Further investigations will be important to screen various developmental stages during metamorphosis and of juvenile animals in order to find the initial steps of eye development and subsequently carry out expression analysis of genes involved in the early genetic cascade of eye development.

In contrast to the adult eye of *Platynereis*, *Pax6* expression was detected in all stages of eye development in the larval eye. Quite similar to the larval eye of *Platynereis*, *Pecten* was found to have larval eyes composed of one pigment cell and a photoreceptor cell (Hodgson and Burke, 1988). The close phylogenetic relationship between *Platynereis* and *Pecten* and the finding of *Pax6* expression in the larval eyes of *Platynereis* suggested a similar role of *Pax6* in the larval eyes of *Pecten*. It is very likely, that most of the lophotrochozoans have larvae with eye spots. Unfortunately, developmental stages are not easily accessible as already mentioned.

Although we focused on pediveliger stages where the eyes start to appear, no *Pax6* expression was detected in the larval eye. Since we were limited to just a few larval stages, it is difficult to interpret this result. Indeed, it can not be ruled out that the proper time-point of *Pax6* expression required for eye development was missed.

Rather unexpected, *Pax6* expression was found in the gill anlagen of *Pecten* larvae. Consistent with this result, *Pax6* expression was detected in the gills of adult animals by quantitative real-time PCR. At present, it can only be speculated, but it probably indicates a putative function of *Pax6* in gill development.

It is noteworthy that bivalves have a paired sensory structure, called osphradium, located along the most lateral margins of the gill axis as raised ridges of tissue (Beninger and Donval, 1995). They consist of both, ciliated sensory cells and bipolar neurons. Osphradia are proposed to have appeared early in the evolution of molluscs (Yonge, 1947) and are most highly developed in gastropods. Initially, a sensory role in monitoring water quality was suggested (Yonge, 1947). Further studies indicated a chemoreceptive function of osphradia (Dorsett, 1986). Moreover, it has been suggested that the osphradia of bivalves may play a role in the reception of chemical spawning cues and the synchronization of gamete release (Beninger and Donval, 1995; Haszprunar, 1987a).

A putative function of *Pax6* in the osphradia of bivalves is supported by the finding that *Pax6* is not only involved in eye morphogenesis but also in the development of the nose and other chemosensory organs (Walther and Gruss, 1991). Consistent with this, *Pax6* expression in cephalopods was found in the olfactory organs and around the suckers on the ventral surface of the arms, which are rich in chemo- and mechanosensory neurons (Hartmann et al., 2003; Tomarev et al., 1997). Interestingly, *Pax6* was also found to be expressed in the gills of the cephalopod *Euprymna* (Hartmann, 2000). The usage of Pax6 in both, the olfactory and visual system supports the idea that these structures may have a common ancestral origin. Notably, there are several common features shared between the development of the olfactory and visual system. In vertebrates, for example, both systems derive from the ectoderm and undergo inductive interactions with defined regions of the brain. Moreover, both placodes share the expression of *Pax6*, *Six3*, *Eya1* and *Dach1* (Chen et al., 1997; Pignoni et al., 1997; Purcell et al., 2005).

Further experiments will aim to determine the precise localization and function of *Pax6* in the gills of bivalves. Moreover, it will be interesting to investigate the evolutionary relationship between the osphradium of bivalves and other olfactory systems.

In a functional assay, both, *AnPax6* and *PmaPax6* were able to induce ectopic eye structures in *Drosophila*. This suggests that at least some of the Pax6 target sequence are conserved between Bivalvia and the fly. It can not be excluded that ectopic expression of bivalvian *Pax6* may lead to the induction of the resident *eyeless* gene. However, the high degree of sequence conservation of the DNA-binding domains suggests the possibility that also other downstream target genes may be regulated.

3. Cloning and expression of AnSix1/2 and PmaSix1/2

3.1 Sequence conservation of AnSix1/2 and PmaSix1/2

We showed that the isolated six genes from *Arca* and *Pecten* belong to the *sine oculis*/Six1/2 subclass. This is corroborated by the diagnostic amino acid sequence ETSY from positions 3 to 6 in helix I of the homeodomain (Seo et al., 1999). In addition, phylogenetic analysis shows clustering of *AnSix1/2* and *PmaSix1/2* to other *sine oculis*/Six1/2 subclass members at high bootstrap values. The six domain of *AnSix1/2* and *PmaSix1/2* show high amino acid sequence identity to the six domain of the polychaete *Platynereis* (94% and 93%, respectively) and more than 90% sequence identity was found to the six domains of vertebrate Six1/2 subclass members. The homeodomain of AnSix1/2 shows highest identity to the homeodomain of *Platynereis* (97%), whereas the homeodomain of PmaSix1/2 shows highest sequence similarity to the homeodomain of Drosophila *sine oculis* (95% sequence identity).

The Six family can be grouped into three major subgroups (Seo et al., 1999). In vertebrates, due to several whole genome duplications, two gene members or more have been identified for each of the three subgroups. The presence of a *so* homolog in sponges (Bebenek et al., 2004) suggests that the Six genes may have arisen very early in metazoan evolution. In cnidarians, which form the closest out-group to the Bilateria, three six genes, one of each subgroup, were isolated (Stierwald et al., 2004). Therefore, it was supposed that the Six gene family arose before the separation of Urbilateria and Cnidaria (Bebenek et al., 2004).

So far, only few studies have been carried out on the Six gene family in lophotrochozoans. However, representatives of each subclass of the Six family genes have been found in planarians (Bebenek et al., 2004; Pineda et al., 2000; Pineda and Salo, 2002). In molluscs, only six genes of the Six1/2 subclass have been isolated so far (Bebenek et al., 2004) but none of the two other subgroups. The recovery of all Six gene subclasses in cnidarians and in basal lophotrochozoans (planarians) strongly suggests that all three subclasses may also exist in molluscs. Further investigations of Six family genes in molluscs are required to understand the conserved function of this gene family.

3.2 Expression of *AnSix1/2* and *PmaSix1/2*

Similar to the situation described for *Arca* and *Pecten Pax6*, neither *Six1/2* expression was detected in the adult compound eyes of *Arca* nor in the adult mirror eyes of *Pecten* by in situ hybridization. Weak *Six1/2* expression was found by real-time PCR, however the high C_T-values indicated by the amplification plot suggest expression below detectable levels for *in situ* hybridization. Although difficult to interpret, these negative results suggest that *Six1/2* may not be involved in the ongoing growth of adult eyes.

In *Drosophila so* is required for the formation of the entire visual system and is expressed from early embryonic stages onwards (Cheyette et al., 1994; Seimiya and Gehring,

2000). Similarly, in the polychaete *Platynereis*, *Six1/2* is continuously expressed in the developing larval and adult eyes (Arendt et al., 2004). In planarians, *Six1/2* is also expressed in regenerating eyes and knock down experiments by RNA interference during regeneration leads to a complete inhibition of eye formation (Pineda et al., 2000).

Due to the fact that *Six1/2* seems to be required for the entire visual system, including the larval eyes of basal metazoans, we were interested whether we can find *Six1/2* expression in the larval eyes of *Pecten*.

In contrast to our initial assumption no *Six1/2* expression was detected in the larval eyes of *Pecten* by *in situ* hybridization. Again, this result is difficult to interpret since we were limited to only a few larval stages and might have missed the essential developmental stages when *Six1/2* is expressed. Interestingly enough, *Six1/2* was found in three distinct regions outside of the eye region. Because of the weak staining signal and the difficulties to identify superimposed structures through the shell, it was not possible to assign the stainings to the corresponding tissues with absolute certainty. However, based on the position of the staining and by comparison with the available anatomical data of *Pecten* larvae, expression is supposed in the gill primordia and in the developing anterior and posterior adductor muscles (personal communication, P. Benninger). Consistent with a putative role of *PmaSix1/2* in muscle and gill development is the finding that genes of the *Six1/2* subfamily are involved in the development of the olfactory system as well as in myogenesis (Laclef et al., 2003b; Oliver et al., 1995a; Zheng et al., 2003). In mice, *Six1* is strongly expressed in the nasal placode and later in the olfactory epithelium, and homozygous mutants for *Six1* display nasal disorganization (Laclef et al., 2003b). Moreover, *Six1* was shown to be required in myogenesis of vertebrates (Heanue et al., 1999; Laclef et al., 2003a). *Six1* is suggested to regulate *MyoD* and *myogenin* expression in the limb buds and $Six1^{-/-}$ homozygous mice die at birth and display severe muscle hypoplasia (Laclef et al., 2003a). Similarly, *Six1/2* expression is also found in the striated muscle layers of jellyfish suggesting a conserved function in muscle development among eumetazoa (Stierwald et al., 2004).

As already mentioned, the gills of bivalves exhibit a sensory field involved in chemoreception. Therefore, it is tempting to speculate that *PmaSix1/2* may have a function in the development of the chemosensory system of the gills as it could be the case for *PmaPax6*. Interestingly, a similar situation is found in vertebrates where both *Pax6* and *Six1* participate in the development of the nose.

4. Cloning and expression of three opsin genes in bivalvian molluscs

In search for opsin genes in bivalves, one opsin gene from *Arca* and two opsin genes from *Pecten* were isolated. The two-dimensional models and the conservation of amino acid residues diagnostic for opsin proteins along with phylogenetic analysis strongly suggest a photosensory role for these genes.

PmaGqOpsin was found to be strongly expressed in the rhabdomeric photoreceptor cells of the proximal retina in the *Pecten* mirror eye. PmaGqOpsin shows high amino acid sequence identity to the Gq-coupled rhodopsin (*Scop1*) isolated from the pacific scallop *Patinopecten yessoensis* (Kojima et al., 1997), belonging to the rhabdomeric-opsin type. Although no expression analysis was carried out for *Scop1*, *in situ* hybridization staining was shown for the α-subunit of the G_q type G-protein, the protein supposed to interact with *Scop1*, in the proximal retina of the mirror eye (Kojima et al., 1997).

Our expression analysis of *PmaGqOpsin* indicates that this rhabdomeric opsin is indeed expressed in the rhabdomeric photoreceptor cells of the *Pecten* mirror eye. The orthology of *PmaGqOpsin* to rhabdomeric opsins is further supported by comparative sequence analysis. At the transition from transmembrane domain VII to the cytoplasmic tail a highly conserved stretch of amino acids (ALSHPKF) was found in PmaGqOpsin, a motif which is specific for rhabdomeric opsins. This stretch was found to play a crucial role in the binding of G-proteins (Marin et al., 2000). Furthermore, the third cytoplasmic loop which is also important to make contact with the G-protein is, as typical for rhabdomeric opsins, considerably longer than in ciliary opsins. Moreover, the counterion for the protonated Schiff base is a tyrosine in *PmaGqOpsin*, another feature shared with all rhabdomeric opsins (Arendt et al., 2004). Consistent with the sequence comparison, phylogenetic analysis shows clustering of *PmaGqOpsin* with rhabdomeric opsins. Comparative sequence and phylogenetic analysis, therefore, univocally identify PmaGqOpsin as a rhabdomeric-opsin type.

The situation is less clear for the second opsin gene isolated from *Pecten*, *PmaOpsinX*. Comparative sequence analysis shows several deviations from other known opsin genes. Within the regions known to be important for interaction with G-proteins, PmaOpsinX shows no sequence homology to any other known opsins. In particular, no sequence homology was found at the transition from the transmembrane domain VII to the cytoplasmic tail where, at least in the case of rhabdomeric and ciliary opsins, a highly conserved amino acid motif is found (SHPK(F/Y)R and VFMNKQF, respectively (Arendt et al., 2004). Furthermore, the third cytoplasmic loop of PmaOpsinX, which is also known to be crucial for G-protein interactions (Terakita et al., 2002) does not show any sequence homology to other opsins.

A peculiarity of *PmaOpsinX* is the usage of a histidine residue as a counterion for the Schiffbase, in contrast to ciliary opsins and rhabdomeric opsins which generally have a glutamate or a tyrosine residue as a counterion. However, variations at this position is not untypical for opsin genes and do not allow any conclusion for the protein function.

Phylogenetic analysis indicates that *PmaOpsinX* most likely clusters to the so far uncharacterized mosquito opsin gene *grop11*. Little support was found that *PmaOpsinX* belongs to any other of the known opsin subfamilies. Interestingly, it is very likely that *PmaOpsinX* may constitute a novel opsin subfamily (34% Local bootstrap probability).

Real-time PCR experiments showed expression in all examined tissues. However, the high C_T-values indicated by the real-time PCR amplification plot and the fact that no *PmaOpsinX* expression was detected by *in situ* hybridization suggests very low expression levels.

Our observation by real-time PCR that *PmaOpsinX* is expressed in various tissues raised the question about its function. Recently, a similar broad expression was found for a novel opsin gene family identified in teleost fish (Moutsaki et al., 2003). These opsin genes, named teleost multiple tissue (tmt) opsins were found to be expressed in a wide variety of tissues (e.g., eyes, kidney, heart, embryonic cell lines). Based on the observation that these opsin genes are expressed in cell lines that possess a light entrainable clock, it was suggested that these tmt-opsins may play an important role for the photic regulation of teleost peripheral circadian clocks (Moutsaki et al., 2003). Therefore, it is tempting to speculate that *PmaOpsinX* acts as a photopigment involved in the photic regulation of peripheral clocks. Further studies will be warranted to search for genes involved in circadian rhythm, as for example the *bmal* gene (an ortholog of the Drosophila *cycle* gene) (Van Gelder, 2004) and to investigate whether expression will co-localize to *PmaOpsinX* expression domains.

The third opsin gene, *AnOpsinX*, was isolated from *Arca*. Comparative sequence analysis and phylogenetic analysis revealed that *AnOpsinX* most probably belongs to the Go-coupled opsin subfamily. At the border between the 7th transmembrane domain and the C-terminal, a stretch of three amino acids is highly indicative of ciliary and rhabdomeric opsin families (Arendt et al., 2004). This amino acid stretch plays an important role in G-protein interaction and alteration of this tripeptide was found to diminish G-protein interaction (Marin et al., 2000). Although it was reported that Go-coupled protein miss such a fingerprint, the isolated *AnOpsinX* shares a SSK motif (position 322, 323 and 324) with two Go-coupled opsin genes recently identified in the sea urchin *Strongylocentrotus* (Raible et al., 2006). The third cytoplasmic loop, however, which is also important for G-protein activation (Terakita et al., 2002) is longer than usual in *AnOpsinX* and shows no sequence homology to any other opsin gene at the corresponding region.

The usually highly conserved E(D)RY motif at the C-terminal end of helix III deviates in *AnOpsinX*. Instead of a glutamate or an aspartate residue at the first position of this motif, *AnOpsinX* has a cysteine (C130). Interestingly, the same substitution is also found in the Go-coupled opsin (*Scop2*) of the pacific scallop *Patinopecten* (Kojima et al., 1997).

Phylogenetic analysis also indicates that *AnOpsinX* most probably clusters to the Go-coupled opsins of Amphioxus (Koyanagi et al., 2002). Moreover, it can be excluded that *AnOpsinX* clusters to the ciliary-opsin, rhabdomeric-opsin or peropsin subfamilies. Recent results from sea urchin (Raible et al., 2006) suggests that the Go-coupled opsins identified in amphioxus, sea urchin and *Patinopecten* comprise an ancient Go-opsin family, supporting the notion that *AnOpsinX* is a member of the Go-coupled opsin subfamily. However, it can not be ruled out (13.8% local bootstrap probability) that *AnOpsinX* may constitute a novel opsin subfamily.

The real-time PCR expression data suggest expression of *AnOpsinX* in the muscle and the gills, whereas very little expression was detected in the eyes and mantle tissue. The finding that *AnOpsinX* is expressed in muscle and gill tissues points to the possibility that *AnOpsinX* may play a role in the regulation of peripheral circadian clocks as already proposed for the putative function of *PmaOpsinX*.

5. Conclusions and Perspectives

The aim of this thesis was to test the recently proposed idea that all eyes found in eumetazoans derive from a common *Pax6*-dependent ancestor eye (Gehring and Ikeo, 1999). The investigations were focused on two bivalvian species, *Arca noae* and *Pecten maximus*, each representing a different eye-type. The fact that different eye-types are present within the same phylogenetic class, makes them perfect candidates to investigate eye evolution.

Two transcription factors were isolated, *Pax6* and *Six1/2*, known to be high up in the regulatory genetic cascade of eye development, from both species. In contrast to the initial assumption, no definite *Pax6* and *Six1/2* expression could be observed in the larval and adult visual systems of *Arca* and *Pecten*. Indeed, in eyes of adult animal only small amounts of

Pax6 and Six1/2 were detected, consistent with the absence of in situ staining in the larval and adult eyes.

These are negative results and therefore difficult to interpret. It is tempting to speculate that both, *Pax6* and *Six1/2,* are probably not necessary for the ongoing growth and maintenance of adult eyes. However, it can not be excluded that *Pax6* and *Six1/2* may be critical in early stages of eye development.

A serious problem of our investigations was the inaccessibility of various embryonic stages of both species. Further studies will aim to investigate a broader window of embryonic stages. An alternative to study larval eyes would be to use an other species easier to breed and to cultivate in the laboratory, as for instance oysters. Although adult animals of oysters have no eyes, the veliger larval stages are known to have eye spots. (Coon et al., 1990; Galtsoff, 1964). Therefore, at least for the investigation of larval eyes, oysters might be the animal of choice.

Surprisingly, *Pax6* expression and very likely also *Six1/2* expression were found in the gill primordium of *Pecten* veliger larvae. Moreover, *Pax6* expression was found in the gills of adult animals in both, *Arca* and *Pecten* by real-time PCR. Although very speculative, *Pax6* and *Six1/2* might be involved in the development of the osphradium. An additional role of *Pax6* could be the maintenance of the sensory components of the osphradium in the adults.

In a second project, three opsin genes were isolated, one from *Arca* and two opsin genes from *Pecten*. Whereas from *Pecten* one opsin gene (*PmaGqOpsin*) was unambiguously identified as a Go-coupled opsin, which is specifically expressed in the rhabdomeric photoreceptor cells of the proximal retina, the classification of the two other opsin genes remains unclear. Phylogenetic analysis indicate that each of them may constitute a novel type of opsin subfamily.

Interestingly, real-time PCR revealed opsin expression in various tissues, suggesting a putative role in the photic regulation of peripheral molecular clocks. To determine whether these opsin genes indeed play a role in the photic regulation of circadian rhythms, it will be useful to search for components of the clocks such as *bmal/cycle*, *per/tim* and to elucidate whether expression of these genes colocalize with opsin expression.

In conclusion, this work provides new molecular comparative insights into the eye evolution of two bivalvian species, each representing a different eye-type. In addition, *Pax6* and *Six1/2* were found to be expressed in the gill primordium of *Pecten* larvae as well as in the adult gills of both *Arca* and *Pecten*. These new findings raise several interesting questions concerning their functional role in the gills and sensory organs in general. Are *Pax6* and

Six1/2 implicated in the establishment of the chemosensory fields of the gills and what is their functional role in adult gills?

Furthermore, two new opsin genes were isolated, which probably constitute novel opsin subfamilies and, strikingly, are expressed in various tissues. These findings opens the door for several more investigations concerning their function and may lead to a better understanding of opsin evolution and the diversification of photoreceptor cells.

VI References

Adachi, J. and Hasegawa, M. (1996). MOLPHY version 2.3: programs for molecular phylogenetics based on maximum likelihood., (ed. Institute of Statistical Mathematics, Tokyo.

Arendt, D. (2003). Evolution of eyes and photoreceptor cell types. *Int J Dev Biol* **47**, 563-71.

Arendt, D., Tessmar-Raible, K., Snyman, H., Dorresteijn, A. W. and Wittbrodt, J. (2004). Ciliary photoreceptors with a vertebrate-type opsin in an invertebrate brain. *Science* **306**, 869-71.

Arendt, D., Tessmar, K., de Campos-Baptista, M. I., Dorresteijn, A. and Wittbrodt, J. (2002). Development of pigment-cup eyes in the polychaete Platynereis dumerilii and evolutionary conservation of larval eyes in Bilateria. *Development* **129**, 1143-54.

Arendt, D. and Wittbrodt, J. (2001). Reconstructing the eyes of Urbilateria. *Philos Trans R Soc Lond B Biol Sci* **356**, 1545-63.

Arshavsky, V. Y., Lamb, T. D. and Pugh, E. N., Jr. (2002). G proteins and phototransduction. *Annu Rev Physiol* **64**, 153-87.

Ashery-Padan, R., Marquardt, T., Zhou, X. and Gruss, P. (2000). Pax6 activity in the lens primordium is required for lens formation and for correct placement of a single retina in the eye. *Genes Dev* **14**, 2701-11.

Barber, V. C., Evans, E. M. and Land, M. F. (1967). The fine structure of the eye of the mollusc Pecten maximus. *Z Zellforsch Mikrosk Anat* **76**, 25-312.

Baumer, N., Marquardt, T., Stoykova, A., Spieler, D., Treichel, D., Ashery-Padan, R. and Gruss, P. (2003). Retinal pigmented epithelium determination requires the redundant activities of Pax2 and Pax6. *Development* **130**, 2903-15.

Beaumont, A. R., Tserpes, G. and M.D., B. (1987). Some effects of copper on the veliger larvae of the mussel *Mytilus edulis* and the scallop *Pecten maximus* (Mollusca, Bivalvia). *Mar. Env. Res.* **21**, 299-309.

Bebenek, I. G., Gates, R. D., Morris, J., Hartenstein, V. and Jacobs, D. K. (2004). sine oculis in basal Metazoa. *Dev Genes Evol* **214**, 342-51.

Becerro, M. A., Uriz, M. J. and Turon, X. (1994). Trends in space occupation by the encrusting sponge *Crambe crambe:* variation in shape with size and environment. *Marine Biology* **121**, 301-307.

Bedini, C., Ferroro, E. and Lanfranchi, A. (1973). Fine structure of the eyes in two species of Dalyelliidae (Turbellaria, Rhabdocoela). *Monit Zool Ital* **7**.

Bellolio, G., Lohrmann, K. and Dupré, E. (1993). Larval morphology of the scallop *Argopecten purpuratus* as revealed by scanning electon microscopy. *Veliger* **36**, 332-342.

Beninger, P. G. and Donval, A. (1995). The ospradium in *Placopecten megallanicus* and *Pecten maximus* (Bivalvia, Pectinidae): histology, ultrastructure, and implications for spawning synchronisation. *Marine Biology* **123**, 121-129.

Beninger, P. G., Dwiono, S. A. P. and Le Pennec, M. (1994). Early development of the gill and implications for feeding in *Pecten maximus* (Bivalvia: Pectinidae). *Marine Biology* **119**, 405-412.

Benović, A. (1997). The hystory, present condition, and future of the molluscan fisheries of Croatia. In *The History, Present Condition, and Future of the Molluscan Fisheries of North and Central America and Europe.*, vol. 3 (ed. C. L. MacKenzie Jr. J. Burrell, V.G. A. Rosenfield and W. L. Hobart), pp. 217-226: NOAA Technical Report NMFS, vol. 129. U.S. Department of Commerce.

Berg, O. and Strand O. (2001). Great scallop, *Pecten maximus*, research and culture strategies in Norway: a review. *Aquaculture International* **9**, 305-318

Blackshaw, S. and Snyder, S. H. (1999). Encephalopsin: a novel mammalian extraretinal opsin discretely localized in the brain. *J Neurosci* **19**, 3681-90.
Blanco, J., Girard, F., Kamachi, Y., Kondoh, H., Gehring, W.J. (2005) Functional analysis of the chicken delta1-crystallin enhancer activity in Drosophila reveals remarkable evolutionary conservation between chicken and fly. *Development* **132**, 1895-905.
Bopp, D., Burri, M., Baumgartner, S., Frigerio, G. and Noll, M. (1986). Conservation of a large protein domain in the segmentation gene paired and in functionally related genes of Drosophila. *Cell* **47**, 1033-40.
Boucher, C. A., Carey, N., Edwards, Y. H., Siciliano, M. J. and Johnson, K. J. (1996). Cloning of the human SIX1 gene and its assignment to chromosome 14. *Genomics* **33**, 140-2.
Boucher, C. A., King, S. K., Carey, N., Krahe, R., Winchester, C. L., Rahman, S., Creavin, T., Meghji, P., Bailey, M. E., Chartier, F. L. et al. (1995). A novel homeodomain-encoding gene is associated with a large CpG island interrupted by the myotonic dystrophy unstable (CTG)n repeat. *Hum Mol Genet* **4**, 1919-25.
Bovolenta, P., Mallamaci, A., Puelles, L. and Boncinelli, E. (1998). Expression pattern of cSix3, a member of the Six/sine oculis family of transcription factors. *Mech Dev* **70**, 201-3.
Bower, S. M. and Meyer, G. R. (1990). Atlas of anatomy and histology of larvae and early juvenile stages of the Japanese scallop (*Patinopecten yessoensis*). *Can. Spec. Publ. Fish. Aquatic Sci.* **111**, 1-51.
Brand, A. H. and Perrimon, N. (1993). Targeted gene expression as a means of altering cell fates and generating dominant phenotypes. *Development* **118**, 401-15.
Breitling, R. and Gerber, J. K. (2000). Origin of the paired domain. *Dev Genes Evol* **210**, 644-50.
Brown, L. S. (2004). Fungal rhodopsins and opsin-related proteins: eukaryotic homologues of bacteriorhodopsin with unknown functions. *Photochem Photobiol Sci* **3**, 555-65.
Cai, J., Lan, Y., Appel, L. F. and Weir, M. (1994). Dissection of the Drosophila paired protein: functional requirements for conserved motifs. *Mech Dev* **47**, 139-50.
Callaerts, P., Halder, G. and Gehring, W. J. (1997). PAX-6 in development and evolution. *Annu Rev Neurosci* **20**, 483-532.
Callaerts, P., Leng, S., Clements, J., Benassayag, C., Cribbs, D., Kang, Y. Y., Walldorf, U., Fischbach, K. F. and Strauss, R. (2001). Drosophila Pax-6/eyeless is essential for normal adult brain structure and function. *J Neurobiol* **46**, 73-88.
Callaerts, P., Munoz-Marmol, A. M., Glardon, S., Castillo, E., Sun, H., Li, W. H., Gehring, W. J. and Salo, E. (1999). Isolation and expression of a Pax-6 gene in the regenerating and intact Planarian Dugesia(G)tigrina. *Proc Natl Acad Sci U S A* **96**, 558-63.
Cano, J. and Garcia, T. (1985). Scallop fishering in the coast of Malaga, S.E. In *Spain 5th Pectinid Workshop, La Coruna, Spain.*, (ed., pp. 8p (mimeo).
Casse, N., Devauchelle, N. and Le Pennec, M. (1998). Embryonic shell formation in the scallop *Pecten maximus* (Linnaeus). *Veliger* **41**, 133-141.
Chalepakis, G., Fritsch, R., Fickenscher, H., Deutsch, U., Goulding, M. and Gruss, P. (1991). The molecular basis of the undulated/Pax-1 mutation. *Cell* **66**, 873-84.
Charman, W. N. (1991). The vertebrate dioptric apparatus. In *Vision and Visual Dysfunction*, vol. 2 (ed. J. F. Cronly-Dillon and R. L. Gregory), pp. 82-117. Basingstoke: Macmillan.
Chen, R., Amoui, M., Zhang, Z. and Mardon, G. (1997). Dachshund and eyes absent proteins form a complex and function synergistically to induce ectopic eye development in Drosophila. *Cell* **91**, 893-903.
Cheyette, B. N., Green, P. J., Martin, K., Garren, H., Hartenstein, V. and Zipursky, S. L. (1994). The Drosophila sine oculis locus encodes a homeodomain-containing protein required for the development of the entire visual system. *Neuron* **12**, 977-96.

Chow, R. L., Altmann, C. R., Lang, R. A. and Hemmati-Brivanlou, A. (1999). Pax6 induces ectopic eyes in a vertebrate. *Development* **126**, 4213-22.

Collinson, J. M., Hill, R. E. and West, J. D. (2000). Different roles for Pax6 in the optic vesicle and facial epithelium mediate early morphogenesis of the murine eye. *Development* **127**, 945-56.

Comely, C. A. (1972). Larval culture of the scallop *Pecten maximus* (L.). *Rapport et Procès-verbaux du Conseil International pour l'Exploration de la Mer* **34**, 365-378.

Coon, S. L., Fitt, W. K. and Bonar, D. B. (1990). Competency and delay of metamorphosis in the Pacific oyster, *Crassostrea gigas* (Thunberg). *Mar. Biol.* **106**, 379-387.

Cragg, S. M. (1985). The adductor and retractor muscles of the veliger of *Pecten maximus* (L.) (Bivalvia). *J. Moll. Stud.* **51**, 276-283.

Czerny, T. and Busslinger, M. (1995). DNA-binding and transactivation properties of Pax-6: three amino acids in the paired domain are responsible for the different sequence recognition of Pax-6 and BSAP (Pax-5). *Mol Cell Biol* **15**, 2858-71.

Czerny, T., Halder, G., Kloter, U., Souabni, A., Gehring, W. J. and Busslinger, M. (1999). twin of eyeless, a second Pax-6 gene of Drosophila, acts upstream of eyeless in the control of eye development. *Mol Cell* **3**, 297-307.

Czerny, T., Schaffner, G. and Busslinger, M. (1993). DNA sequence recognition by Pax proteins: bipartite structure of the paired domain and its binding site. *Genes Dev* **7**, 2048-61.

Dakin, W. J. (1910). The eye of *Pecten*. *Q. J. Microsc. Sci.* **340**, 49-112.

Dorsett, D. A. (1986). Brains to cells: the neuroanatomy of selected gastropod species. In *The Mollusca, Vol 9. Neurobiology and behaviour, Part 2.*, (ed. A. O. D. Willows), pp. 101-187. Orlando: Academic press.

Dozier, C., Kagoshima, H., Niklaus, G., Cassata, G. and Burglin, T. R. (2001). The Caenorhabditis elegans Six/sine oculis class homeobox gene ceh-32 is required for head morphogenesis. *Dev Biol* **236**, 289-303.

Drew, G. A. (1906). The habits, anatomy and embryology of the giant scallop (*Pecten tenuicostatus*, Mighels). *Univ. Maine Stud.* **6**, 77 pp. + 17 plates.

Eakin, R. M. (1963). Lines of evolution of photoreceptors. In *General physiology of cell specialization.*, (ed. D. Mazia and A. Tyler), pp. 393-425. New York: McGraw-Hill.

Eakin, R. M. (1968). Evolution of photoreceptors. New York: Appleton-Century-Crofts.

Eakin, R. M. (1982). Continuity and diversity in photoreceptors. In *Visual cells in Evolution.*, (ed. J. A. Westfall). New York: Raven Press.

Eberhard, D., Jimenez, G., Heavey, B. and Busslinger, M. (2000). Transcriptional repression by Pax5 (BSAP) through interaction with corepressors of the Groucho family. *Embo J* **19**, 2292-303.

Ellison-Wright, Z., Heyman, I., Frampton, I., Rubia, K., Chitnis, X., Ellison-Wright, I., Williams, S. C., Suckling, J., Simmons, A. and Bullmore, E. (2004). Heterozygous PAX6 mutation, adult brain structure and fronto-striato-thalamic function in a human family. *Eur J Neurosci* **19**, 1505-12.

Epstein, J., Cai, J., Glaser, T., Jepeal, L. and Maas, R. (1994a). Identification of a Pax paired domain recognition sequence and evidence for DNA-dependent conformational changes. *J Biol Chem* **269**, 8355-61.

Epstein, J. A., Glaser, T., Cai, J., Jepeal, L., Walton, D. S. and Maas, R. L. (1994b). Two independent and interactive DNA-binding subdomains of the Pax6 paired domain are regulated by alternative splicing. *Genes Dev* **8**, 2022-34.

Fernald, R. D. (2006). Casting a genetic light on the evolution of eyes. *Science* **313**, 1914-8.

Fu, W. and Noll, M. (1997). The Pax2 homolog sparkling is required for development of cone and pigment cells in the Drosophila eye. *Genes Dev* **11**, 2066-78.

Fujiwara, M., Uchida, T., Osumi-Yamashita, N. and Eto, K. (1994). Uchida rat (rSey): a new mutant rat with craniofacial abnormalities resembling those of the mouse Sey mutant. *Differentiation* **57**, 31-8.
Fullarton, J. H. (1896). On the development of the common scallop (*Pecten opercularis* L.). *Rep. Fish. Board Scot.* **6**, 290-299.
Furuta, Y. and Hogan, B. L. (1998). BMP4 is essential for lens induction in the mouse embryo. *Genes Dev* **12**, 3764-75.
Gaines, P. and Carlson, J. R. (1995). The olfactory and visual systems are closely related in Drosophila. *Braz J Med Biol Res* **28**, 161-7.
Galtsoff, P. S. (1964). The American Oyster, *Crassostrea virginica* Gmelin. *Fish. Bull.* **64**, 1-480.
Gehring, W. and Rosbash, M. (2003). The coevolution of blue-light photoreception and circadian rhythms. *J Mol Evol* **57 Suppl 1**, S286-9.
Gehring, W. J. (2004). Historical perspective on the development and evolution of eyes and photoreceptors. *Int J Dev Biol* **48**, 707-17.
Gehring, W. J. and Ikeo, K. (1999). Pax 6: mastering eye morphogenesis and eye evolution. *Trends Genet* **15**, 371-7.
Gérard, M. (1995). *PAX*-genes expression during human embryonic development, a preliminary report. *C. R. Acad. Sci. Paris*, 57-66.
Glardon, S., Callaerts, P., Halder, G. and Gehring, W. J. (1997). Conservation of Pax-6 in a lower chordate, the ascidian Phallusia mammillata. *Development* **124**, 817-25.
Glardon, S., Holland, L. Z., Gehring, W. J. and Holland, N. D. (1998). Isolation and developmental expression of the amphioxus Pax-6 gene (AmphiPax-6): insights into eye and photoreceptor evolution. *Development* **125**, 2701-10.
Glaser, T., Jepeal, L., Edwards, J. G., Young, S. R., Favor, J. and Maas, R. L. (1994). PAX6 gene dosage effect in a family with congenital cataracts, aniridia, anophthalmia and central nervous system defects. *Nat Genet* **7**, 463-71.
Gorman, A. L. and McReynolds, J. S. (1969). Hyperpolarizing and depolarizing receptor potentials in the scallop eye. *Science* **165**, 309-10.
Grifone, R., Demignon, J., Houbron, C., Souil, E., Niro, C., Seller, M. J., Hamard, G. and Maire, P. (2005). Six1 and Six4 homeoproteins are required for Pax3 and Mrf expression during myogenesis in the mouse embryo. *Development* **132**, 2235-49.
Grindley, J. C., Davidson, D. R. and Hill, R. E. (1995). The role of Pax-6 in eye and nasal development. *Development* **121**, 1433-42.
Gruffyd, L. D. and Beaumont, A. R. (1970). Determination of the optimum concentration of eggs and spermatozoa for the production of normal larvae in Pecten maximus (Mollusca, Lamellibranchia). *Helgol. wiss. Meeresunts.* **20**, 486-497.
Gruffyd, L. D. and Beaumont, A. R. (1972). A method of rearing *Pecten maximus* larvae in the laboratory. *Marine Biology* **2**, 64-70.
Gutsell, J. S. (1930). Natural history of the Bay scallop. *Bull. U.S. Bur. Fish.* **46**, 569-632.
Halder, G., Callaerts, P. and Gehring, W. J. (1995). Induction of ectopic eyes by targeted expression of the eyeless gene in Drosophila. *Science* **267**, 1788-92.
Hamm, H. E. and Gilchrist, A. (1996). Heterotrimeric G proteins. *Curr Opin Cell Biol* **8**, 189-96.
Hao, W., Chen, P. and Fong, H. K. (2000). Analysis of chromophore of RGR: retinal G-protein-coupled receptor from pigment epithelium. *Methods Enzymol* **316**, 413-22.
Hardie, R. C. (2001a). Phototransduction in Drosophila melanogaster. *J Exp Biol* **204**, 3403-9.
Hardie, R. C. and Raghu, P. (2001b). Visual transduction in Drosophila. *Nature* **413**, 186-93.

Hargrave, P. A. and McDowell, J. H. (1992). Rhodopsin and phototransduction: a model system for G protein-linked receptors. *Faseb J* **6**, 2323-31.
Harris, W. A. (1997). Pax-6: where to be conserved is not conservative. *Proc Natl Acad Sci U S A* **94**, 2098-100.
Hartline, H. K. (1938). The discharge of impulses in the optic nerve of *Pecten* in response to illumination of the eye. *J. Cell. Comp. Physiol.* **11**, 465-478.
Hartmann, B. (2000). Cloning of a *Pax-6* homologue in the cephalopod *Euprymna scolopes* and characterization of its expression pattern during embryonic development. In *Fakultät für Biologie*, (ed. Freiburg in Breisgau: Albert-Ludwig-Universität.
Hartmann, B., Lee, P. N., Kang, Y. Y., Tomarev, S., de Couet, H. G. and Callaerts, P. (2003). Pax6 in the sepiolid squid Euprymna scolopes: evidence for a role in eye, sensory organ and brain development. *Mech Dev* **120**, 177-83.
Hartmann, E., Rapoport, T. A. and Lodish, H. F. (1989). Predicting the orientation of eukaryotic membrane-spanning proteins. *Proc Natl Acad Sci U S A* **86**, 5786-90.
Haszprunar, G. (1987a). The fine morphology of the osphradial sense organs of Mollusca. III. Placophora and Bivalvia. *Phil. Trans. R. Soc. Lond. (Ser B)* **315**, 37-61.
Haszprunar, G. (1987b). The fine morphology of the osphradial sense organs of the Mollusca. IV. Caudofoveata and Solenogastres. *Phil. Trans. R. Soc. Lond. (Ser B)* **315**, 63-73.
Hauck, B., Gehring, W. J. and Walldorf, U. (1999). Functional analysis of an eye specific enhancer of the eyeless gene in Drosophila. *Proc Natl Acad Sci U S A* **96**, 564-9.
Heanue, T. A., Reshef, R., Davis, R. J., Mardon, G., Oliver, G., Tomarev, S., Lassar, A. B. and Tabin, C. J. (1999). Synergistic regulation of vertebrate muscle development by Dach2, Eya2, and Six1, homologs of genes required for Drosophila eye formation. *Genes Dev* **13**, 3231-43.
Hesse, R. (1897). Untersuchungen über die Organe der Lichtempfindung bei niederen Thieren. II. Die Augen der Plathelminthen. *Z. wiss. Zool.* **62**, 527-582.
Hill, M. E., Asa, S. L. and Drucker, D. J. (1999). Essential requirement for Pax6 in control of enteroendocrine proglucagon gene transcription. *Mol Endocrinol* **13**, 1474-86.
Hill, R. E., Favor, J., Hogan, B. L., Ton, C. C., Saunders, G. F., Hanson, I. M., Prosser, J., Jordan, T., Hastie, N. D. and van Heyningen, V. (1991). Mouse small eye results from mutations in a paired-like homeobox-containing gene. *Nature* **354**, 522-5.
Hittner, H. M. (1989). Aniridia. In *The Glaucomas*, pp. 869-884. St. Louis: The C.V. Mosby Comp.
Hodgson, C. A. and Burke, R. D. (1988). Development and larval morphology of the spiny scallop, *Chlamys hastata*. *Bio. Bull.* **174**, 303-318.
Hoge, M. A. (1915). Another gene in the fourth chromosome of *Drosophila*. *Am. Nat.* **49**, 47-49.
Hoshiyama, D., Suga, H., Iwabe, N., Koyanagi, M., Nikoh, N., Kuma, K., Matsuda, F., Honjo, T. and Miyata, T. (1998). Sponge Pax cDNA related to Pax-2/5/8 and ancient gene duplications in the Pax family. *J Mol Evol* **47**, 640-8.
Hrs-Brenko, M. and Legac, M. (1996). A review of bivalve species in the eastern Adriatic Sea: II. Pteriomorphia (Arcidae and Noetidae). *Nat. Croat.* **5**, 221-247.
Jagger, W. S. (1992). The optics of the spherical fish lens. *Vision Res* **32**, 1271-84.
Janssen, H. H. (1991). Die rätselhaften Augen der antarktischen Muschel *Lissarca notocarcensis*. *Mikrokosmos* **80**, 109-112.
Jares-Erijman, E. A., Sakai, R. and Rinehart, K. L. (1991). Crambescidins: new antiviral and cytotoxic compounds form the sponge *Crambe crambe*. *Journal of Organic Chemistry* **56**, 5712-5715.
Johns, P. R. (1977). Growth of the adult goldfish eye. III. Source of the new retinal cells. *J Comp Neurol* **176**, 343-57.

Kamachi, Y., Cheah, K. S. and Kondoh, H. (1999). Mechanism of regulatory target selection by the SOX high-mobility-group domain proteins as revealed by comparison of SOX1/2/3 and SOX9. *Mol Cell Biol* **19**, 107-20.

Kamachi, Y., Uchikawa, M., Tanouchi, A., Sekido, R. and Kondoh, H. (2001). Pax6 and SOX2 form a co-DNA-binding partner complex that regulates initiation of lens development. *Genes Dev* **15**, 1272-86.

Kammermeier, L., Leemans, R., Hirth, F., Flister, S., Wenger, U., Walldorf, U., Gehring, W. J. and Reichert, H. (2001). Differential expression and function of the Drosophila Pax6 genes eyeless and twin of eyeless in embryonic central nervous system development. *Mech Dev* **103**, 71-8.

Kawakami, K., Ohto, H., Takizawa, T. and Saito, T. (1996). Identification and expression of six family genes in mouse retina. *FEBS Lett* **393**, 259-63.

Kawamura, S. and Yokoyama, S. (1998). Functional characterization of visual and nonvisual pigments of American chameleon (Anolis carolinensis). *Vision Res* **38**, 37-44.

Kirby, R. J., Hamilton, G. M., Finnegan, D. J., Johnson, K. J. and Jarman, A. P. (2001). Drosophila homolog of the myotonic dystrophy-associated gene, SIX5, is required for muscle and gonad development. *Curr Biol* **11**, 1044-9.

Kirschfeld, K. (1976). The resolution of lens and compound eyes. In *Neural principles in vision*, (ed. F. Zettler and R. Weiler), pp. 354-370. Berlin: Springer.

Klein, P. S., Sun, T. J., Saxe, C. L., 3rd, Kimmel, A. R., Johnson, R. L. and Devreotes, P. N. (1988). A chemoattractant receptor controls development in Dictyostelium discoideum. *Science* **241**, 1467-72.

Kleinjan, D. A., Seawright, A., Schedl, A., Quinlan, R. A., Danes, S. and van Heyningen, V. (2001). Aniridia-associated translocations, DNase hypersensitivity, sequence comparison and transgenic analysis redefine the functional domain of PAX6. *Hum Mol Genet* **10**, 2049-59.

Kobayashi, M., Toyama, R., Takeda, H., Dawid, I. B. and Kawakami, K. (1998). Overexpression of the forebrain-specific homeobox gene six3 induces rostral forebrain enlargement in zebrafish. *Development* **125**, 2973-82.

Kojima, D. and Fukada, Y. (1999). Non-visual photoreception by a variety of vertebrate opsins. *Novartis Found Symp* **224**, 265-79; discussion 279-82.

Kojima, D., Terakita, A., Ishikawa, T., Tsukahara, Y., Maeda, A. and Shichida, Y. (1997). A novel Go-mediated phototransduction cascade in scallop visual cells. *J Biol Chem* **272**, 22979-82.

Korf, H. W. (1994). The pineal organ as a component of the biological clock. Phylogenetic and ontogenetic considerations. *Ann N Y Acad Sci* **719**, 13-42.

Koyanagi, M., Terakita, A., Kubokawa, K. and Shichida, Y. (2002). Amphioxus homologs of Go-coupled rhodopsin and peropsin having 11-cis- and all-trans-retinals as their chromophores. *FEBS Lett* **531**, 525-8.

Kozmik, Z. (2005). Pax genes in eye development and evolution. *Curr Opin Genet Dev* **15**, 430-8.

Kozmik, Z., Czerny, T. and Busslinger, M. (1997). Alternatively spliced insertions in the paired domain restrict the DNA sequence specificity of Pax6 and Pax8. *Embo J* **16**, 6793-803.

Kozmik, Z., Daube, M., Frei, E., Norman, B., Kos, L., Dishaw, L. J., Noll, M. and Piatigorsky, J. (2003). Role of Pax genes in eye evolution: a cnidarian PaxB gene uniting Pax2 and Pax6 functions. *Dev Cell* **5**, 773-85.

Krauss, S., Johansen, T., Korzh, V., Moens, U., Ericson, J. U. and Fjose, A. (1991). Zebrafish pax[zf-a]: a paired box-containing gene expressed in the neural tube. *Embo J* **10**, 3609-19.

Kronhamn, J., Frei, E., Daube, M., Jiao, R., Shi, Y., Noll, M. and Rasmuson-Lestander, A. (2002). Headless flies produced by mutations in the paralogous Pax6 genes eyeless and twin of eyeless. *Development* **129**, 1015-26.
Kuchiiwa, T., Kuchiiwa, S. and Teshirogi, W. (1991). Comparative morphological studies on the visual system in a binocular and a multi-ocular species of freshwater planarian. *Hydrobiologia* **227**, 241-249.
Kulikova, V. A. and Tabunkov, V. D. (1974). Ecology, reproduction, growth and productive properties of a population of the scallop *Mizuchopecten yessoensis* Dysodonta, Pectinidae in the Busset Lagoon. *Aniva Bay. Zool. Zh.* **53**, 1767-1774.
Laclef, C., Hamard, G., Demignon, J., Souil, E., Houbron, C. and Maire, P. (2003a). Altered myogenesis in Six1-deficient mice. *Development* **130**, 2239-52.
Laclef, C., Souil, E., Demignon, J. and Maire, P. (2003b). Thymus, kidney and craniofacial abnormalities in Six 1 deficient mice. *Mech Dev* **120**, 669-79.
Lamb, T. D. (1996). Gain and kinetics of activation in the G-protein cascade of phototransduction. *Proc Natl Acad Sci U S A* **93**, 566-70.
Land, M. F. (1965). Image formation by a concave reflector in the eye of the scallop, Pecten maximus. *J Physiol* **179**, 138-53.
Land, M. F. (1966). Activity in the optic nerve of Pecten maximus in response to changes in light intensity, and to pattern and movement in the optical environment. *J Exp Biol* **45**, 83-99.
Land, M. F. (1984). Crustacea. In *Photoreception and Vision in Invertebrates*, (ed. M. A. Ali), pp. 401-438. New York: Plenum.
Land, M. F. and Nilsson, D.-E. (2002). Animal eyes. Oxford: Oxford University Press.
Lang, D., Lu, M. M., Huang, L., Engleka, K. A., Zhang, M., Chu, E. Y., Lipner, S., Skoultchi, A., Millar, S. E. and Epstein, J. A. (2005). Pax3 functions at a nodal point in melanocyte stem cell differentiation. *Nature* **433**, 884-7.
Lauderdale, J. D., Wilensky, J. S., Oliver, E. R., Walton, D. S. and Glaser, T. (2000). 3' deletions cause aniridia by preventing PAX6 gene expression. *Proc Natl Acad Sci U S A* **97**, 13755-9.
Le Pennec, M. (1974). Morphogenèse de la coquille de *Pecten maximus* (L.) élevé au laboratoire. *Cahier de Biologie Marine* **15**, 475-482.
Le Pennec, M., Paugaum, A. and Le Pennec, G. (2003). The pelagic life of the pectinid *Pecten maximus*-a review. *ICES Journal of Marine Science* **60**, 211-223.
Levi, P. and Levi, C. (1971). Ultrastructure des yeux palleaux d'*Arca noae* (L.). *J. Microsc. (Paris)* **11**, 425-432.
Li, X., Oghi, K. A., Zhang, J., Krones, A., Bush, K. T., Glass, C. K., Nigam, S. K., Aggarwal, A. K., Maas, R., Rose, D. W. et al. (2003). Eya protein phosphatase activity regulates Six1-Dach-Eya transcriptional effects in mammalian organogenesis. *Nature* **426**, 247-54.
Loosli, F., Kmita-Cunisse, M. and Gehring, W. J. (1996). Isolation of a Pax-6 homolog from the ribbonworm Lineus sanguineus. *Proc Natl Acad Sci U S A* **93**, 2658-63.
Loosli, F., Winkler, S. and Wittbrodt, J. (1999). Six3 overexpression initiates the formation of ectopic retina. *Genes Dev* **13**, 649-54.
Macdonald, R. and Wilson, S. W. (1997). Distribution of Pax6 protein during eye development suggest discrete roles in proliferation and differentiated cells. *Development Genes and Evolution* **206**, 363-369.
Malakhov, V. V. and Medvedeva, L. A. (1986). Embryonic development in the bivalves *Patinopecten yessoensis* (Pectinida, Pectinidae) and *Sipsula sachalinensis* (Cardiida, Mactridae). *Zool. Zh.* **65**, 732-740.
Marin, A. and Lopez Belluga, M. D. (2004). Sponge coating decrease predation on the bivalve *Arca noae*. *J. Moll. Stud.* **71**, 1-6.

Marin, E. P., Krishna, A. G., Zvyaga, T. A., Isele, J., Siebert, F. and Sakmar, T. P. (2000). The amino terminus of the fourth cytoplasmic loop of rhodopsin modulates rhodopsin-transducin interaction. *J Biol Chem* **275**, 1930-6.

Marquardt, T., Ashery-Padan, R., Andrejewski, N., Scardigli, R., Guillemot, F. and Gruss, P. (2001). Pax6 is required for the multipotent state of retinal progenitor cells. *Cell* **105**, 43-55.

Martinez-Morales, J. R., Rodrigo, I. and Bovolenta, P. (2004). Eye development: a view from the retina pigmented epithelium. *Bioessays* **26**, 766-77.

Maru, K. (1972). Morphological observations on the veliger larvae of the scallop *Patinopecten yessoensis* (Jay). *Sci. Rep. Hokkaido Fish. Expl. Stn.* **14**, 55-62.

Mason, J. (1983). Scallop and queen fisheries in the British Isles.

Miki, N., Keirns, J. J., Marcus, F. R., Freeman, J. and Bitensky, M. W. (1973). Regulation of cyclic nucleotide concentrations in photoreceptors: an ATP-dependent stimulation of cyclic nucleotide phosphodiesterase by light. *Proc Natl Acad Sci U S A* **70**, 3820-4.

Miller, W. H. (1958). Derivatives of cilia in the distal sense cells of the retina of Pecten. *J Biophys Biochem Cytol* **4**, 227-8.

Minchin, D. (1978). An exceptionally large escallop (*Pecten maximus* (L.)) from west Cork. *Irish Naturalist's Journal* **19**, 202.

Moore, R. Y. and Lenn, N. J. (1972). A retinohypothalamic projection in the rat. *J Comp Neurol* **146**, 1-14.

Morton, B. (1987). The pallial photophores of *Barbatia virescens* (Bivalvia: Arcacea). *J. Moll. Stud.* **53**, 241-243.

Moutsaki, P., Whitmore, D., Bellingham, J., Sakamoto, K., David-Gray, Z. K. and Foster, R. G. (2003). Teleost multiple tissue (tmt) opsin: a candidate photopigment regulating the peripheral clocks of zebrafish? *Brain Res Mol Brain Res* **112**, 135-45.

Muntz, W. R. A. and Raj, U. (1984). On the visual system of *Nautilus pompilius. J. Exp. Biol.* **109**, 253-263.

Neethirajan, G., Krishnadas, S. R., Vijayalakshmi, P., Shashikant, S. and Sundaresan, P. (2004). PAX6 gene variations associated with aniridia in south India. *BMC Med Genet* **5**, 9.

Nelson, L. B., Spaeth, G. L., Nowinski, T. S., Margo, C. E. and Jackson, L. (1984). Aniridia. A review. *Surv Ophthalmol* **28**, 621-42.

Nelson, R. J. and Zucker, I. (1981). Photoperiodic control of reproduction in olfactory-bulbectomized rats. *Neuroendocrinology* **32**, 266-71.

Niimi, T., Seimiya, M., Kloter, U., Flister, S. and Gehring, W. J. (1999). Direct regulatory interaction of the eyeless protein with an eye-specific enhancer in the sine oculis gene during eye induction in Drosophila. *Development* **126**, 2253-60.

Nilsson, D.-E. (1989). Optics and evolution of the compound eye. In *Facets of vision*, (ed. D. G. Stavenga and R. C. Hardie), pp. 30-73. Berlin: Springer.

Nilsson, D.-E. (1994). Eyes as optical alarm systems in fan worms and ark clams. *Phil. Trans. R. Soc. Lond. (Ser B)*, 195-212.

Nilsson, D. E. (2004). Eye evolution: a question of genetic promiscuity. *Curr Opin Neurobiol* **14**, 407-14.

Noll, M. (1993). Evolution and role of Pax genes. *Curr Opin Genet Dev* **3**, 595-605.

Nordsieck, F. (1969). Die europäischen Meeresmuscheln (Bivalvia) vom Eismeer bis Kapverden, Mittelmeer und Schwarzes Meer. Stuttgart: Gustav Fisher Verlag.

Nordstrom, K., Wallen, R., Seymour, J. and Nilsson, D.-E. (2003). A simple visual system without neurons in jellyfish larvae. *Proc Biol Sci* **270**, 2349-54.

Noveen, A., Daniel, A. and Hartenstein, V. (2000). Early development of the Drosophila mushroom body: the roles of eyeless and dachshund. *Development* **127**, 3475-88.
Oliver, G., Loosli, F., Koster, R., Wittbrodt, J. and Gruss, P. (1996). Ectopic lens induction in fish in response to the murine homeobox gene Six3. *Mech Dev* **60**, 233-9.
Oliver, G., Mailhos, A., Wehr, R., Copeland, N. G., Jenkins, N. A. and Gruss, P. (1995b). Six3, a murine homologue of the sine oculis gene, demarcates the most anterior border of the developing neural plate and is expressed during eye development. *Development* **121**, 4045-55.
Oliver, G., Wehr, R., Jenkins, N. A., Copeland, N. G., Cheyette, B. N., Hartenstein, V., Zipursky, S. L. and Gruss, P. (1995a). Homeobox genes and connective tissue patterning. *Development* **121**, 693-705.
Onuma, Y., Takahashi, S., Asashima, M., Kurata, S. and Gehring, W. J. (2002). Conservation of Pax 6 function and upstream activation by Notch signaling in eye development of frogs and flies. *Proc Natl Acad Sci U S A* **99**, 2020-5.
Palczewski, K., Kumasaka, T., Hori, T., Behnke, C. A., Motoshima, H., Fox, B. A., Le Trong, I., Teller, D. C., Okada, T., Stenkamp, R. E. et al. (2000). Crystal structure of rhodopsin: A G protein-coupled receptor. *Science* **289**, 739-45.
Patten, W. (1886). Eyes of molluscs and arthropods. *Mitt. Zool. Statz. Neapel* **6**, 542-756.
Pauli, T., Seimiya, M., Blanco, J. and Gehring, W. J. (2005). Identification of functional sine oculis motifs in the autoregulatory element of its own gene, in the eyeless enhancer and in the signalling gene hedgehog. *Development* **132**, 2771-82.
Peharda, M., Mladineo, I., Bolotin, J., Kekez, L., Skaramuca, B. (2006). The reproductive cycle and potential protantric development of the Noah's Ark shell, *Arca noae* L.: Implications for aquaculture. *Aquaculture* **252**, 317-327
Perron, M., Kanekar, S., Vetter, M. L. and Harris, W. A. (1998). The genetic sequence of retinal development in the ciliary margin of the Xenopus eye. *Dev Biol* **199**, 185-200.
Piatigorsky, J. and Kozmik, Z. (2004). Cubozoan jellyfish: an Evo/Devo model for eyes and other sensory systems. *Int J Dev Biol* **48**, 719-29.
Piccinetti, C., Simunovic, A. and Jukic, S. (1986). Distribution and abundance of *Chlamys opercularis* (L.) and *Pecten jacobaeus* L. in the Adriatic Sea. *FAO Fisheries Report No. 345, FIPL/R345*, pp. 99-105.
Pichaud, F. and Desplan, C. (2002). Pax genes and eye organogenesis. *Curr Opin Genet Dev* **12**, 430-4.
Pignoni, F., Hu, B., Zavitz, K. H., Xiao, J., Garrity, P. A. and Zipursky, S. L. (1997). The eye-specification proteins So and Eya form a complex and regulate multiple steps in Drosophila eye development. *Cell* **91**, 881-91.
Pineda, D., Gonzalez, J., Callaerts, P., Ikeo, K., Gehring, W. J. and Salo, E. (2000). Searching for the prototypic eye genetic network: Sine oculis is essential for eye regeneration in planarians. *Proc Natl Acad Sci U S A* **97**, 4525-9.
Pineda, D. and Salo, E. (2002). Planarian Gtsix3, a member of the Six/so gene family, is expressed in brain branches but not in eye cells. *Mech Dev* **119 Suppl 1**, S167-71.
Plaza, S., Prince, F., Jaeger, J., Kloter, U., Flister, S., Benassayag, C., Cribbs, D. and Gehring, W. J. (2001). Molecular basis for the inhibition of Drosophila eye development by Antennapedia. *Embo J* **20**, 802-11.
Poppe, G. T. and Goto, Y. (2000). European seashells. Scaphopoda, Bivalvia, Cephalopoda, vol II. Hackenheim: Conchbooks.
Prosser, J. and van Heyningen, V. (1998). PAX6 mutations reviewed. *Hum Mutat* **11**, 93-108.
Provencio, I., Jiang, G., De Grip, W. J., Hayes, W. P. and Rollag, M. D. (1998). Melanopsin: An opsin in melanophores, brain, and eye. *Proc Natl Acad Sci U S A* **95**, 340-5.

Provencio, I., Rodriguez, I. R., Jiang, G., Hayes, W. P., Moreira, E. F. and Rollag, M. D. (2000). A novel human opsin in the inner retina. *J Neurosci* **20**, 600-5.
Punzo, C., Seimiya, M., Flister, S., Gehring, W. J. and Plaza, S. (2002). Differential interactions of eyeless and twin of eyeless with the sine oculis enhancer. *Development* **129**, 625-34.
Purcell, P., Oliver, G., Mardon, G., Donner, A. L. and Maas, R. L. (2005). Pax6-dependence of Six3, Eya1 and Dach1 expression during lens and nasal placode induction. *Gene Expr Patterns* **6**, 110-8.
Puschel, A. W., Gruss, P. and Westerfield, M. (1992). Sequence and expression pattern of pax-6 are highly conserved between zebrafish and mice. *Development* **114**, 643-51.
Quiring, R., Walldorf, U., Kloter, U. and Gehring, W. J. (1994). Homology of the eyeless gene of Drosophila to the Small eye gene in mice and Aniridia in humans. *Science* **265**, 785-9.
Raible, F., Tessmar-Raible, K., Arboleda, E., Kaller, T., Bork, P., Arendt, D. and Arnone, M. I. (2006). Opsins and clusters of sensory G-protein-coupled receptors in the sea urchin genome. *Dev Biol*.
Ralph, M. R., Foster, R. G., Davis, F. C. and Menaker, M. (1990). Transplanted suprachiasmatic nucleus determines circadian period. *Science* **247**, 975-8.
Rhode, K. and Watson, N. A. (1991). Ultrastructure of pigmented photoreceptors of larval *Multicotyle purvisi* (Trematoda, Aspidogastrea). *Parasito. Res.* **77**, 485-490.
Richardson, J., Cvekl, A. and Wistow, G. (1995). Pax-6 is essential for lens-specific expression of zeta-crystallin. *Proc Natl Acad Sci U S A* **92**, 4676-80.
Robert, R. and Gérard, A. (1999). Bivalve hatchery technology: The current situation for the Pacific oyster *Crassostrea gigas* and the scallop *Pecten maximus* in France. *Aquat. Living Resour.* **12**, 121-130.
Salvini-Plawen, L. and Mayr, E. (1961). In *Evolutionary Biology*, vol. 10 (ed. M. K. Hecht W. C. Steere and B. Wallace), pp. pp. 207-263: Plenum Press.
Sarkar, P. S., Appukuttan, B., Han, J., Ito, Y., Ai, C., Tsai, W., Chai, Y., Stout, J. T. and Reddy, S. (2000). Heterozygous loss of Six5 in mice is sufficient to cause ocular cataracts. *Nat Genet* **25**, 110-4.
Sarkar, P. S., Paul, S., Han, J. and Reddy, S. (2004). Six5 is required for spermatogenic cell survival and spermiogenesis. *Hum Mol Genet* **13**, 1421-31.
Sastry, A. N. (1965). The development and external morphology of pelagic larval and post-larval stages of the bay scallop *Aequipecten irradians concentricus* Say, reared in the laboratory. *Bull. Mar. Sci. Gulf Caribb.* **15**, 417-435.
Seimiya, M. and Gehring, W. J. (2000). The Drosophila homeobox gene optix is capable of inducing ectopic eyes by an eyeless-independent mechanism. *Development* **127**, 1879-86.
Seo, H. C., Curtiss, J., Mlodzik, M. and Fjose, A. (1999). Six class homeobox genes in drosophila belong to three distinct families and are involved in head development. *Mech Dev* **83**, 127-39.
Seo, H. C., Drivenes, O., Ellingsen, S. and Fjose, A. (1998). Transient expression of a novel Six3-related zebrafish gene during gastrulation and eye formation. *Gene* **216**, 39-46.
Serikaku, M. A. and O'Tousa, J. E. (1994). sine oculis is a homeobox gene required for Drosophila visual system development. *Genetics* **138**, 1137-50.
Sisodiya, S. M., Free, S. L., Williamson, K. A., Mitchell, T. N., Willis, C., Stevens, J. M., Kendall, B. E., Shorvon, S. D., Hanson, I. M., Moore, A. T. et al. (2001). PAX6 haploinsufficiency causes cerebral malformation and olfactory dysfunction in humans. *Nat Genet* **28**, 214-6.
Sivak, J. G., West, J. A. and Campbell, M. C. (1994). Growth and optical development of the ocular lens of the squid (Sepioteuthis lessoniana). *Vision Res* **34**, 2177-87.

Sollars, P. J., Smeraski, C. A., Kaufman, J. D., Ogilvie, M. D., Provencio, I. and Pickard, G. E. (2003). Melanopsin and non-melanopsin expressing retinal ganglion cells innervate the hypothalamic suprachiasmatic nucleus. *Vis Neurosci* **20**, 601-10.
Spudich, J. L., Yang, C. S., Jung, K. H. and Spudich, E. N. (2000). Retinylidene proteins: structures and functions from archaea to humans. *Annu Rev Cell Dev Biol* **16**, 365-92.
St-Onge, L., Sosa-Pineda, B., Chowdhury, K., Mansouri, A. and Gruss, P. (1997). Pax6 is required for differentiation of glucagon-producing alpha-cells in mouse pancreas. *Nature* **387**, 406-9.
Stierwald, M., Yanze, N., Bamert, R. P., Kammermeier, L. and Schmid, V. (2004). The Sine oculis/Six class family of homeobox genes in jellyfish with and without eyes: development and eye regeneration. *Dev Biol* **274**, 70-81.
Straznicky, K. and Gaze, R. M. (1971). The growth of the retina in Xenopus laevis: an autoradiographic study. *J Embryol Exp Morphol* **26**, 67-79.
Sun, H., Dickinson, D. P., Costello, J. and Li, W. H. (2001). Isolation of Cladonema Pax-B genes and studies of the DNA-binding properties of cnidarian Pax paired domains. *Mol Biol Evol* **18**, 1905-18.
Sun, H., Gilbert, D. J., Copeland, N. G., Jenkins, N. A. and Nathans, J. (1997). Peropsin, a novel visual pigment-like protein located in the apical microvilli of the retinal pigment epithelium. *Proc Natl Acad Sci U S A* **94**, 9893-8.
Tanaka, Y. (1984). On the development of *Pecten albicans* Schroeter. *Bull. Natl. Res. Inst. Aquacult. (Japan)* **5**, 19-25.
Tarttelin, E. E., Bellingham, J., Hankins, M. W., Foster, R. G. and Lucas, R. J. (2003). Neuropsin (Opn5): a novel opsin identified in mammalian neural tissue. *FEBS Lett* **554**, 410-6.
Tebble, N. (1966). British bivalve seashells. *British Museum, London*, 212.
Terakita, A., Yamashita, T., Nimbari, N., Kojima, D. and Shichida, Y. (2002). Functional interaction between bovine rhodopsin and G protein transducin. *J Biol Chem* **277**, 40-6.
Tomarev, S. I., Callaerts, P., Kos, L., Zinovieva, R., Halder, G., Gehring, W. and Piatigorsky, J. (1997). Squid Pax-6 and eye development. *Proc Natl Acad Sci U S A* **94**, 2421-6.
Ton, C. C., Hirvonen, H., Miwa, H., Weil, M. M., Monaghan, P., Jordan, T., van Heyningen, V., Hastie, N. D., Meijers-Heijboer, H., Drechsler, M. et al. (1991). Positional cloning and characterization of a paired box- and homeobox-containing gene from the aniridia region. *Cell* **67**, 1059-74.
Ton, C. C., Miwa, H. and Saunders, G. F. (1992). Small eye (Sey): cloning and characterization of the murine homolog of the human aniridia gene. *Genomics* **13**, 251-6.
Tootle, T. L., Silver, S. J., Davies, E. L., Newman, V., Latek, R. R., Mills, I. A., Selengut, J. D., Parlikar, B. E. and Rebay, I. (2003). The transcription factor Eyes absent is a protein tyrosine phosphatase. *Nature* **426**, 299-302.
Toy, J., Yang, J. M., Leppert, G. S. and Sundin, O. H. (1998). The optx2 homeobox gene is expressed in early precursors of the eye and activates retina-specific genes. *Proc Natl Acad Sci U S A* **95**, 10643-8.
Treisman, J., Harris, E. and Desplan, C. (1991). The paired box encodes a second DNA-binding domain in the paired homeo domain protein. *Genes Dev* **5**, 594-604.
Tsonis, P. A. and Fuentes, E. J. (2006). Focus on molecules: Pax-6, the eye master. *Exp Eye Res* **83**, 233-4.
Underhill, D. A. and Gros, P. (1997). The paired-domain regulates DNA binding by the homeodomain within the intact Pax-3 protein. *J Biol Chem* **272**, 14175-82.
Van Gelder, R. N. (2004). Recent insights into mammalian circadian rhythms. *Sleep* **27**, 166-71.

van Heyningen, V. and Williamson, K. A. (2002). PAX6 in sensory development. *Hum Mol Genet* **11**, 1161-7.
Vogan, K. J., Underhill, D. A. and Gros, P. (1996). An alternative splicing event in the Pax-3 paired domain identifies the linker region as a key determinant of paired domain DNA-binding activity. *Mol Cell Biol* **16**, 6677-86.
Waller, T. R. (1980). Scanning electron microscopy of shell and mantle in the order Arcoida, Mollusca, Bivalvia. *Smithson. Contrib. Zool.* **313**, 1-58.
Walther, C. and Gruss, P. (1991). Pax-6, a murine paired box gene, is expressed in the developing CNS. *Development* **113**, 1435-49.
Wawersik, S., Purcell, P., Rauchman, M., Dudley, A. T., Robertson, E. J. and Maas, R. (1999). BMP7 acts in murine lens placode development. *Dev Biol* **207**, 176-88.
Wetts, R. and Fraser, S. E. (1988). Multipotent precursors can give rise to all major cell types of the frog retina. *Science* **239**, 1142-5.
Wetts, R., Serbedzija, G. N. and Fraser, S. E. (1989). Cell lineage analysis reveals multipotent precursors in the ciliary margin of the frog retina. *Dev Biol* **136**, 254-63.
Wilson, D. S., Guenther, B., Desplan, C. and Kuriyan, J. (1995). High resolution crystal structure of a paired (Pax) class cooperative homeodomain dimer on DNA. *Cell* **82**, 709-19.
Xu, W., Rould, M. A., Jun, S., Desplan, C. and Pabo, C. O. (1995). Crystal structure of a paired domain-DNA complex at 2.5 A resolution reveals structural basis for Pax developmental mutations. *Cell* **80**, 639-50.
Yonge, C. M. (1947). The pallial organs in the aspidobranch Gastropoda and their evolution throughout the Mollusca. *Phil. Trans. R. Soc. Lond. (Ser B)* **232**, 443-518.
Zhang, Y. and Emmons, S. W. (1995). Specification of sense-organ identity by a Caenorhabditis elegans Pax-6 homologue. *Nature* **377**, 55-9.
Zheng, W., Huang, L., Wei, Z. B., Silvius, D., Tang, B. and Xu, P. X. (2003). The role of Six1 in mammalian auditory system development. *Development* **130**, 3989-4000.
Zuber, M. E., Perron, M., Philpott, A., Bang, A. and Harris, W. A. (1999). Giant eyes in Xenopus laevis by overexpression of XOptx2. *Cell* **98**, 341-52.

Die VDM Verlagsservicegesellschaft sucht für wissenschaftliche Verlage abgeschlossene und herausragende

Dissertationen, Habilitationen, Diplomarbeiten, Master Theses, Magisterarbeiten usw.

für die kostenlose Publikation als Fachbuch.

Sie verfügen über eine Arbeit, die hohen inhaltlichen und formalen Ansprüchen genügt, und haben Interesse an einer honorarvergüteten Publikation?

Dann senden Sie bitte erste Informationen über sich und Ihre Arbeit per Email an *info@vdm-vsg.de*.

Sie erhalten kurzfristig unser Feedback!

VDM Verlagsservicegesellschaft mbH
Dudweiler Landstr. 99
D - 66123 Saarbrücken
www.vdm-vsg.de

Telefon +49 681 3720 174
Fax   +49 681 3720 1749

Die VDM Verlagsservicegesellschaft mbH vertritt

Printed by Books on Demand GmbH, Norderstedt / Germany